SAS/GRAPH®
Beyond the Basics

Robert Allison

The correct bibliographic citation for this manual is as follows: Allison, Robert. 2012. *SAS/GRAPH®: Beyond the Basics*. Cary, NC: SAS Institute Inc.

SAS/GRAPH®: Beyond the Basics

Contents

Acknowledgements v

Preface.. vii

Example 1 Bar with Annotated Marker 1

Example 2 Using PROC GREPLAY For Custom Dashboard Layout......................... 7

Example 3 Paneling Multiple Graphs on a Page...17

Example 4 Using an Image As Part of the Graph 27

Example 5 Customizing a Map 33

Example 6 Overlay G3D Surfaces 41

Example 7 Adding Border Space 47

Example 8 Determine Customers Within Map Borders 55

Example 9 Determine Customers Within Circular Radius 63

Example 10 Custom Box Plots 65

Example 11 Using Company Logos in Graphs 71

Example 12 Annotated Map Borders 77

Example 13 Population Tree Charts..87

Example 14 Sparkline Table..97

Example 15 Custom Waterfall Chart 109

Example 16 Plotting Data on Floor Plans 117

Example 17 Custom Calendar Chart.. 127

Example 18 Fancy Line Graph with Broken Axis... 141

Example 19 Drill-down Link to an HTML Anchor... 149

Example 20 Time Series "Strip Plot" 157

Example 21 GIF Animation... 163

Example 22 Using SAS/IntrNet with Graphs 169

Example 23 Plotting Coordinates on a Map... 175

Example 24 Plotting Coverage Areas on a Map 185

Example 25 Plotting Multiple Graphs on the Same Page 193

Example 26 Grand Finale: An Advanced Dashboard 201

Now What 225

Index... 227

Acknowledgements

So, how did I become a SAS/GRAPH expert and end up writing this book? There are many people who played a role in that, but I'd specifically like to thank the following:

Dr. Moon Suh, my advisor at NCSU, who first introduced me to SAS software, encouraged me to become a "data meister," and challenged me to learn how to create analytic graphics.

Dr. David Dickey, my time series professor at NCSU, who "showed me the light" with his tricky SAS examples in class and at user conferences, and showed me that anything is possible with SAS/GRAPH.

Mike Zdeb, who has worked on joint conference papers with me, and who sends me "ideas" and "challenges" on an almost weekly basis so that I don't run out of ideas.

Stephen Few, who gave me the opportunity to win the DM Review dashboard contest and who also allows me to bounce ideas off him. He keeps me pointed in the right direction in pursuit of "graphical excellence."

SAS Institute, for providing SAS software to NCSU (for free), where I used it for my dissertation project. As the saying goes, "When your only tool is a hammer, every problem looks like a nail." SAS software was (and still is) a superb hammer!

Meg Pounds, who was my boss for 10 years, and who gave me the flexibility to spend my testing time experimenting with the SAS/GRAPH software, learning and inventing new things to do with it, and working with the developers to request new features that enable us to create even better custom graphs.

Robert Dolan, who played a major role in adding several of the underlying features that I use such as anti-aliasing, support for more than 16 million colors, and transparency.

Kathleen Ramage, who has continued to support and enhance the traditional SAS/GRAPH procedures such as PROC GCHART, by adding the key features I need in order to create the perfect graph.

And my Mom, who let me draw graphics on the attic walls and encouraged me to design posters for 4-H demonstrations and a slew of other things growing up.

I would also like to thank the following technical reviewers and others who helped out behind the scenes:

Steve England - Technical Reviewer
Marcia Surratt - Technical Reviewer
Mike Zdeb -Technical Reviewer
Brad Kellam - Copy Editor
Marchellina Waugh - Cover Design
Jennifer Dilley - Graphics
Denise Jones - Production Specialist
Stacey Hamilton - Marketing Specialist
Mary Beth Steinbach - Managing Editor
George McDaniel – Development Editor

A wise man once said:

"A motivated SAS/GRAPH programmer can create just about any graph."

In this book, I try to give you the knowledge and inspiration to be that programmer.

These days, anybody can create a simple graph. Many companies provide software with a simple GUI (Graphical User Interface) that enables users to point-and-click their way through a gallery of graphs, select which variables to plot, and produce a mediocre graph.

Given how easy it is to make them, you might ask, "How can I make my graphs stand out?"

A lot of people fall into the trap of trying to make their graphs stand out by being *fancier*. They add bright colors, three-dimensional effects, images in the background, and sometimes make the charts fade-in, spring-up, or oscillate before coming to a rest where you can finally read them. But fancier is not necessarily better, and in the case of graphs, such fancy additions usually make the graph more difficult to read and quickly comprehend, so making the graph fancier can actually make the graph worse rather than better.

I encourage you to make your graphs stand out by being *better*, rather than fancier. And by better, I mean graphs that show the data more clearly, and make it easier for people to look at the graph and quickly answer the questions the graph was created to answer in the first place. And in order to do that, you will need to learn to customize your graphs.

That is where SAS/GRAPH comes in. Whereas most software provides only a GUI interface that enables you to create the graphs that are pre-programmed, SAS/GRAPH provides a programming framework that enables you to customize every aspect of the graphs. It even enables you to create totally new and unique graphs.

Very few people take the time to learn how to create better graphs through customization, and therefore knowing how to do this will give you a great competitive advantage.

Purpose of This Book

The purpose of this book is to go way beyond the basics, and teach you how to create your own custom graphs, not just to modify the look of the graphs, but to create your own graphs with totally unique geometry, layouts, and so on.

Target Audience

The target audience for this book is the SAS/GRAPH programmer who has a firm grasp of the basics and is motivated to learn more.

When I say "programmer," I mean someone who writes SAS programs. The techniques described in this book require programming, and in order to utilize and modify these techniques to produce your own unique custom graphs, you will need to understand the programming tricks.

In addition to SAS/GRAPH programming, you will also need to be comfortable with data manipulation. You will often need to transpose, de-normalize, or otherwise restructure your data to get it to work with the various SAS/GRAPH procedures.

Also, creating clever annotate data sets is a very important part of creating custom graphs. Annotate is a way of letting you programmatically add text or graphics to a graph in SAS. You store the annotation functions in a SAS data set, and each observation in the data set contains one command. For example, "at this X/Y coordinate, do this thing."

Software Used in This Book

First, you will need to be comfortable manipulating data using SAS tools such as the following.

- DATA step

- PROC SQL

- PROC TRANSPOSE

- various functions to manipulate text variables (TRIM, LEFT, SCAN, SUBSTR, and so on)

Of course, you do not need to know everything about those tools, but you will at least need to be well versed with the basics and be comfortable using them.

In addition to those data manipulation tools, you will also need to be comfortable with the following, at least on a basic level:

- using ODS HTML and `device=png` for creating Web output

- macro variable substitution

- running a simple macro and passing a parameter to it (in a few examples)

The examples in this book primarily use the following SAS/GRAPH procedures.

- PROC GPLOT

- PROC GCHART

- PROC GMAP

- PROC GSLIDE

- PROC GREPLAY

Time will be spent explaining how to use PROC GREPLAY (since most users probably have not used it before), but since you should already have a good basic understanding of the other procedures, the basics will not be explained in this book.

Software Not Used in This Book

A single book cannot cover everything, and here is a list of a few things related to SAS/GRAPH that are not covered in this book:

- Java and ActiveX output.

- The SAS 9.2 graphics that use the ODS GRAPHICS ON statement (also referred to as ODS Statistical Graphics or SAS/STAT graphs).

- The SAS 9.2 procedures based on the Graph Template Language (GTL) such as the SGPLOT and SGPANEL procedures.

- Graph procedures that do not produce a GRSEG entry (that is, cannot be used with the GREPLAY procedure).

- SAS/GRAPH procedures that do not support annotation.

Although the Java and ActiveX versions of the SAS/GRAPH graphs are interesting and fun, they do not support all the methods that can be used to create custom graphics. Many of the procedure options are only partially supported, you cannot use PROC GREPLAY with them, and you cannot use the GTITLE and GFOOTNOTE options (placing title text within the graph area—see Example 7). I rely on all of these things for my custom graph "tricks."

The new ODS Statistical Graphics and GTL-based SAS graphs are typically designed to produce nice graphs for a special target audience out-of-the-box, and do not require (or allow) extensive customizations. They allow you to change the "look" of the graphs, but not the geometry itself so that you can come up with new or unique graphs.

And since PROC GREPLAY and the Annotate facility are the major customization techniques described in this book, any graphical procedure that does not support those is not used in these examples. This would include some of the new SAS/GRAPH procedures such as PROC GKPI, PROC GAREABAR, and PROC GTILE.

My One SAS/GRAPH Program

I jokingly tell people that although I have been a SAS programmer for over 18 years, I have written only *one* SAS program: I just keep modifying it. To a large extent, that is actually true.

Here I describe the basic shell I use for all my SAS/GRAPH examples. All the samples in this book are coded using this same basic technique. I always say that there are 10 different ways to do things in SAS. You do not *have* to write your SAS/GRAPH programs this way, but I would strongly encourage you to at least consider these techniques. They have served me well over the years.

I assign a macro variable called NAME at the top of the job, and then I use that as the name of the HTML output file (in the ODS statement), and also as the name of the PNG output file (in the DES= option). I set a filename called "odsout" to "." (the current working directory) and use that as my ODS HTML path so that both the HTML output and the PNG graphics file are written there. Of course, you can use any folder you want, such as "C:\Someplace\" instead of ".", if you prefer.

Note that I also name my .sas job by this same name, so that both the SAS job and the output it produces have the same name; this makes it easy to keep the files organized. For example, the following (pseudo-code) would be stored in a file named foo.sas, and it would produce the output in foo.png and foo.htm:

```
%let name=foo;
filename odsout '.';

goptions device=png;

ODS LISTING CLOSE;
ODS HTML path=odsout body="&name..htm";

proc whatever data=mydata;
 someplot / des="" name="&name";
run;

quit;
ODS HTML CLOSE;
ODS LISTING;
```

As far as the order of things, I generally work with my data first (IMPORT, SUMMARIZE, TRANSPOSE, and so on), and then turn on my ODS statements just before I want to write out my output. This way, I do not have to do things to suppress output when I am manipulating the data, such as use the NOPRINT option in PROC SQL.

These days I find that the most useful and flexible type of output is created using the ODS HTML statement, which produces output that is suitable for viewing on the Web. You can easily view your output with a Web browser (either on your local file system or on a Web server), and you can easily share it with others (they need only a Web browser to view it). Also, you can add tooltips and drill-down functionality to pieces of the graphs, which is very useful.

I specify `device=png` in all my examples. If you are wanting to create custom graphs, I strongly recommend that you do not use `device=java` or `device=activex`. They do not support all the SAS/GRAPH procedure options (see the "partially supported" notes throughout the reference manual), and they do not support PROC GREPLAY at all. I choose PNG over GIF, mainly because PNG supports 16 million colors, whereas the GIF standard has a limit of 256. Also, the SAS cutting-edge new development and new features are focused on PNG rather than GIF.

I typically use GOPTIONS statements to control the font and size of the text, so I can easily control that in one place, and I typically specify `gunit=pct` so I can specify font sizes as a percent of the height of the page. This way, if I make the graphs larger or smaller, the text is automatically resized accordingly. I will occasionally specify the font size in points (such as `htext=10pt`) if I specifically want the text to be that point size, no matter how big or small I make the graph. Specifying point size is sometimes useful for publications that require all text to be in a certain size.

And I generally find that using the GOPTIONS XPIXELS and YPIXELS to control the size of the output is the best way to go. Below is a simple program that demonstrates most of the techniques described above:

```
%let name=prototype_graph_program;

filename odsout '.';

data mydata;
input category $ 1-1 value;
datalines;
A 1
B 2
C 3
D 4
;
run;


goptions device=png;
goptions xpixels=500 ypixels=400;

ODS LISTING CLOSE;
ODS HTML path=odsout body="&name..htm" (title="Prototype Graph")
style=sasweb;

goptions gunit=pct htitle=6 ftitle="albany amt/bo" htext=5
ftext="albany amt";

axis1 label=none;
axis2 label=none minor=none;

pattern v=solid color=cx43A2CA;

title1 "My Prototype SAS/GRAPH Program";

proc gchart data=mydata;
hbar category / type=sum sumvar=value
 nostats noframe
 maxis=axis1
 raxis=axis2
 des="" name="&name";
run;
```

```
quit;
ODS HTML CLOSE;
ODS LISTING;
```

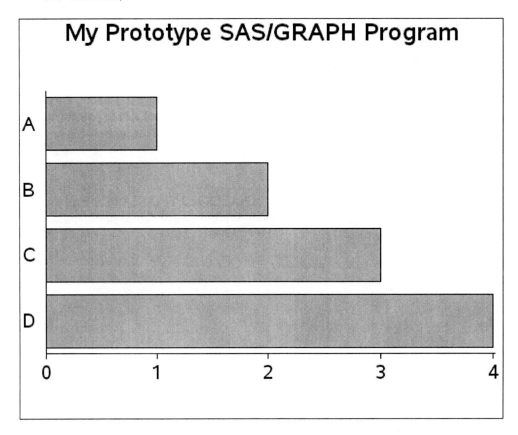

Running the Graph Code

If you are reading an online copy of the book, you could hypothetically copy and paste the code from the book into your SAS editor. But I recommend you download the SAS code in its entirety from the author's Web site (see "Author Pages" below), and run the complete program, rather than trying to run the bits and pieces of code as I try to explain what each piece does. Many of the fragments of code depend on other fragments of code having been run first, and perhaps other macro variables and other settings that were performed earlier in the full code version (but not shown in the book, for brevity). Also, sometimes I do not explain the pieces of code in the exact same order they appear in the actual program.

It might surprise you to hear that I recommend you run your SAS/GRAPH jobs in batch mode, and I strongly recommend that as the best way to develop custom graph code.

You could run your code in the standard interactive DMS (Display Management System) SAS interface, but you would continually have to reset your graphics options, delete your GRSEGS entries, and so on, and also try to decide whether you are looking at the graph from the current run or the previous run. At a minimum, you should probably add the following to the top of all your programs to reset your goptions and delete all the GRSEGS entries if you run them in DMS:

```
goptions reset=all;

%macro delcat(catname);
 %if %sysfunc(cexist(&catname)) %then %do;
  proc greplay nofs igout=&catname;
  delete _all_;
  run;
 %end;
 quit;
%mend delcat;
%delcat(work.gseg);
```

By comparison, when you run each of your jobs in a separate batch run (as I recommend), you are guaranteed a clean start each time, and only the options and settings that you specifically set in your code are the ones that are controlling the appearance of your custom graphs.

You could also run your SAS/GRAPH jobs in SAS Enterprise Guide (EG), but you would first need to remove the hardcoded `ods html` and so on from my sample code. And you would still have similar problems as with DMS SAS, wondering whether the results you see are based on the code you are currently running, or a combination of that and the code you previously ran. EG also tries very hard to make you use ActiveX, colorful ODS styles, and the NOGTITLES option (that is, the titles are separate text outside the graph). Therefore, you will need to do a lot of work to change the EG defaults before you can generate custom graphs.

To run a SAS job in batch, I just make sure the SAS executable is in my search path, **cd** to the directory (or folder) containing the SAS job (such as foo.sas), and then run **sas foo.sas** from the command prompt. The output .htm and .png files are written to the current directory as well as the SAS log.

Here is an example on a PC:

```
Command Prompt                                          _ □ ×

C:\>u:

U:\>cd public_html

U:\public_html>sas foo.sas

U:\public_html>dir foo.*
 Volume in drive U is U31
 Volume Serial Number is 0502-4FD0

 Directory of U:\public_html

10/14/2010  02:16 PM              4,377 foo.log
10/14/2010  02:16 PM              3,641 foo.htm
10/14/2010  02:15 PM              1,403 foo.sas
10/14/2010  02:16 PM             19,034 foo.png
               4 File(s)          28,455 bytes
               0 Dir(s)      161,050,624 bytes free

U:\public_html>
```

Here is an example on UNIX:

```
*****                                                  _ □ ×
% cd ~/public_html
% sas foo.sas
% ls foo.*
foo.htm  foo.log  foo.png  foo.sas
%
```

To write my code, I use whatever editor I want to edit the .sas file (usually the vi editor on UNIX) instead of using the SAS program editor in DMS SAS. And I view the .log file in another window using the 'vi' editor or a file browser. (I do not get the benefit of the DMS SAS color-coding, but I generally find the WARNING and ERROR messages in the log point me in the right direction.)

I edit the .sas file in one window, submit the 'sas foo.sas' in another window (using "command line recall" so I do not have to type the same command over and over), and view the HTML or PNG output in a Web browser. And I iteratively repeat, over and over, until the graph is as close to perfect as it can get.

Just some background information—both my PC and UNIX computers can access the same (multi-protocol) file server and therefore I usually edit my code from UNIX, and submit the code from my PC. I do this because I am most comfortable with the UNIX editor and file management, and because I prefer PC SAS (my PC typically has more fonts that I can use in my SAS jobs, and the pre-production versions of SAS are available on PCs first, since that is the main development platform). I encourage you to choose the platforms that will be most convenient to you.

You do not *have* to edit your SAS jobs outside of DMS SAS, and submit your SAS/GRAPH jobs in batch, but I find it to be a very useful and flexible way to work, especially when writing custom graphs.

Delivering and Viewing Your Output

I strongly encourage you to view your output in a Web browser.

While you can generally view files in your local file system (that is, any file system that your computer can access) by using the file system path in your browser, I recommend using a "proper" Web server, if possible (ideally, on the Web server where your output will be delivered to your users). This will help you make sure the interactivity of your graphs is working, and that your drill-down paths do not contain any absolute pathnames to files on your local drive. Also, putting your output on a real Web server makes it easy to share with others by sending them a URL.

In my case, I have a public_html folder which I can access directly from Windows or UNIX, and everything under that public_html folder can be viewed through the Web server on our intranet via a Web browser.

If you cannot deliver your output to your users via a Web server, one alternative is to E-mail it to them. If your output does not use any HTML tooltips or drill-down functionality, you could simply mail them the .png file. If your output uses HTML tooltips or drill-down functionality, then you will need to mail both the .png and the .htm file, and they will have to save them to the disk, and then view it in their browser.

Author Pages

Each SAS Press author has an author page, which includes several features that relate to the author including a biography, book descriptions for coming soon titles and other titles by the author, contact information, links to sample chapters and example code and data, events and extras, and more. You can access the author pages from http://support.sas.com/authors.

You can access the example programs for this book by linking to http://support.sas.com/publishing/authors/allison_robert.html. Select this book, and then click Example Code and Data.

If you are unable to access the code through the Web site, send e-mail to saspress@sas.com.

Additional Resources

SAS offers you a rich variety of resources to help build your SAS skills and explore and apply the full power of SAS software. Whether you are in a professional or academic setting, we have learning products that can help you maximize your investment in SAS.

Bookstore	http://support.sas.com/publishing/
Training	http://support.sas.com/training/

In the bookstore, you might specifically be interested in "Maps Made Easy Using SAS." Per training courses, you might find the following courses useful: "SAS/GRAPH 1: Essentials" and "Producing Maps with SAS/GRAPH."

Knowledge Base	http://support.sas.com/resources/
Support	http://support.sas.com/techsup/
Learning Center	http://support.sas.com/learn/
Community	http://support.sas.com/community/ sas_graph_and_ods_graphics

Comments or Questions?

If you have comments or questions about this book, you may contact the author through SAS as follows:

Mail:

SAS Institute Inc.
SAS Press
Attn: Robert Allison
SAS Campus Drive
Cary, NC 27513

E-mail: saspress@sas.com

Fax: (919) 677-4444

Please include the title of the book in your correspondence.

For a complete list of books available through SAS Press, visit support.sas.com/publishing.

SAS Publishing News:

Receive up-to-date information about all new SAS publications via e-mail by subscribing to the SAS Publishing News monthly eNewsletter. Visit support.sas.com/subscribe.

EXAMPLE 1

Bar with Annotated Marker

Purpose: Demonstrate how to annotate custom graphic markers onto a chart in a data-driven manner.

Let us start with a simple little bar chart. This bar chart technique was developed as part of the winning entry from SAS/GRAPH in DM Review Magazine's Dashboard Contest.[1] (See Example 2, "Using PROC GREPLAY for Custom Dashboard Layout," for a description of the dashboard in its entirety.)

The data set consists of quarter (Q1-Q4), the actual value attained, and the target value. We will use the following simple values:

```
data metrics;
input Quarter $ 1-2 Actual Target;
datalines;
Q1 3.1833 3.25
Q2 2.9500 3.50
Q3 2.8167 3.75
Q4 1.8126 4.00
;
run;
```

The performance criteria are as follows:

```
    <60% = poor
 60%-90% = satisfactory
   >=90% = good
```

In my quest for the "perfect graph," I started with a simple bar chart, showing just the actual values.

```
axis1 label=none order=(0 to 5 by 1) minor=none major=(h=2)
offset=(0,0);

axis2 label=none;

pattern1 v=s c=grayee;

proc gchart data=metrics;
 vbar quarter / type=sum sumvar=actual
 raxis=axis1 maxis=axis2 coutline=gray width=14 space=14;
run;
```

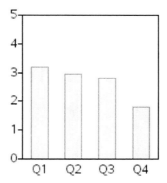

But this graph told only a small part of the story, and was definitely not going to win a dashboard contest. In addition to the actual values, I wanted to show the target value, and also indicate whether performance was poor, satisfactory, or good.

There was no built-in graphing option for what I had in mind, but custom SAS/GRAPH programming enabled me to implement what I had in mind.

Target Line

First I annotated a target line above each bar. I used the MOVE and DRAW Annotate functions to create a line over each bar.

In the annotate data set, first specify xsys='2' and ysys='2' to use the data coordinate system, and set the X value (called a midpoint for vbar charts) to the quarter, and the Y value to the target, and move to that location. Then switch to xsys='7' (relative % coordinate system) and move 8.5% to the left, and from that location draw a line 17% to the right (17% = 2*8.5%). This produces a target line, centered above each bar.

```
data targets; set metrics;
length style color $20 function $8;
hsys='3'; position='5'; when='a';
function='move';
xsys='2'; ysys='2'; midpoint=quarter; y=target; output;
function='move'; xsys='7'; x=-8.5; output;
function='draw'; xsys='7'; x=17; color="gray"; size=.1; output;
run;

proc gchart data=metrics anno=targets;
 vbar quarter / type=sum sumvar=actual
 raxis=axis1 maxis=axis2 coutline=gray width=14 space=14;
run;
```

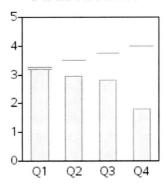

Target Marker

The annotated lines add a reference to visually compare the bars to the target values, but the chart still does not show which performance criteria range the bars are in.

To do the latter I annotated a triangular shaped marker at the end of the target line, and colored it to indicate whether performance was poor or satisfactory or good. I used the A character of the SAS/GRAPH MARKER font for the triangle, but you could use other characters (and other fonts), if desired.

First, you need to measure the actual value against the target:

```
data metrics; set metrics;
format Percent_of_Target percent6.0;
length Evaluation $12;
 percent_of_target=actual/target;
       if (percent_of_target < .60) then evaluation='Poor';
 else if (percent_of_target < .90) then evaluation='Satisfactory';
 else if (percent_of_target >=.90) then evaluation='Good';
run;
```

Defining the colors in macro variables makes it convenient to maintain them in one location and use them in several locations throughout the program.

```
%let green=cxc2e699;
%let pink=cxfa9fb5;
%let red=cxff0000;
```

Create the annotated markers, of the appropriate color, by adding the following lines to the annotated TARGETS data set.

```
data targets; set metrics (where=(target^=.));
length style color $20 function $8;
hsys='3'; position='5'; when='a';
function='move';
xsys='2'; ysys='2'; midpoint=quarter; y=target; output;
function='move'; xsys='7'; x=-8.5; output;
function='draw'; xsys='7'; x=17; color="gray"; size=.1; output;

function='move';
xsys='2'; ysys='2'; midpoint=quarter; y=target; output;
function='move'; xsys='7'; x=8.5; output;
function='cntl2txt'; output;
function='label'; text='A'; size=4; xsys='7'; x=0;
```

```
style='marker';
if (evaluation eq 'Good') then color="&green";
else if (evaluation eq 'Poor') then color="&red";
else if (evaluation eq 'Satisfactory') then color="&pink";
output;
```

If you find a command in these examples that is not familiar to you, you can either accept that it just works, or you can look it up in the SAS Help. For example, if you are curious about the CNTL2TXT annotate function, the Help will tell you that it "copies the values of the internal coordinates stored in the variable pairs (XLAST, YLAST) to (XLSTT, YLSTT)." Which then might lead you to find that (XLAST, YLAST) "tracks the last values specified for the X and Y variables when X and Y are used with non-text functions," and (XLSTT, YLSTT) "tracks the last position for the X and Y variables when X and Y are used with text-handling functions." In other words, you use the CNTL2TXT function when you change from drawing to writing text in the same location where you were drawing.

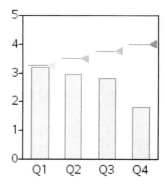

Satisfaction

Note: If you are looking at this publication in black and white, the images here do not show the color. To see the colors look at the color image on the author's Web page.[2]

For a nice touch I added a border around the triangle (specifying the same color as the horizontal line), using the character A from the SAS/GRAPH MARKERE (marker empty) font. This way, if a light color is used against a light background (such as the light green with a white background), the triangle is still easily visible:

```
function='move';
xsys='2'; ysys='2'; midpoint=quarter; y=target; output;
function='move'; xsys='7'; x=8.5; output;
function='cntl2txt'; output;
function='label'; text='A'; size=4; xsys='7'; x=0;
style='markere';
color="gray";
output;
run;

proc gchart data=metrics anno=targets;
 vbar quarter / type=sum sumvar=actual
 raxis=axis1 maxis=axis2 coutline=gray width=14 space=14;
run;
```

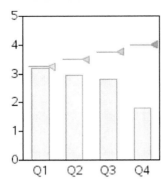

Colored Bars

I wanted to color the bars the same as the triangles. Colors can be easily assigned to bars using PROC GCHART'S SUBGROUP= option in combination with pattern statements. In this case, where the chart has three SUBGROUP= values, they are assigned as Good=pattern1, Poor=pattern2, Satisfactory=pattern3 (the values are assigned to the patterns in alphabetical order).

But what happens if the data changes the next time we generate a graph, and it does not have any bar subgroup values to map to a certain pattern color? That could change the order in which the colors are assigned!

In this graph, the colors hold special meaning—you *always* want green to mean Good, pink to mean Satisfactory, and red to mean Poor. Therefore, to guarantee that the colors are always assigned in that order, you need to insert some data with SAS missing values for each possible value of subgroup.

```
data foometrics;
  length evaluation $12;
  quarter='Q1'; actual=.;
  evaluation='Poor'; output;
  evaluation='Satisfactory'; output;
  evaluation='Good'; output;
run;

data metrics;
  set metrics foometrics;
run;
```

Note that these extra observations with missing numeric values could adversely affect a *frequency* count bar chart, but they have no adverse impact on a bar chart where the height of the bar is the *sum* of the values.

```
pattern1 v=s c=&green;
pattern2 v=s c=&red;
pattern3 v=s c=&pink;

proc gchart data=metrics anno=targets;
  vbar quarter / type=sum sumvar=actual
  subgroup=evaluation nolegend
  raxis=axis1 maxis=axis2 autoref cref=graycc clipref
  coutline=gray width=14 space=14;
run;
```

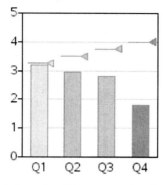

As you can see, this "simple little bar chart" is not so simple!

Notes

[1] http://www.information-management.com/issues/20050901/1035522-1.html.
[2] http://support.sas.com/publishing/authors/allison_robert.html.

EXAMPLE 2

Using PROC GREPLAY For Custom Dashboard Layout

Purpose: Demonstrate how to create a custom SAS/GRAPH GREPLAY template and display multiple graphs in it.

Building on Example 1, "Bar with Annotated Marker," this example combines several of those "simple little bar charts" (and a few other graphics) together onto the same page into what is called an *information dashboard*. The actual code for the entire dashboard (over 1500 lines) can be downloaded from the author's Web site. The code pertinent to creating a custom GREPLAY template and displaying graphs in it is described in detail here.

Overly Simple Dashboards (tiling)

Creating dashboards generally involves placing several graphs on the same page, and the extent to which you can control the exact size, proportions, and placement of those graphs can greatly influence the effectiveness of the dashboard. Some software simply "tiles" several graphs onto the same page (in simple rows and columns), and calls it a dashboard. Although the tiling technique is easy, the dashboards produced using that technique do not live up to the full potential of a well-designed dashboard with a custom layout.

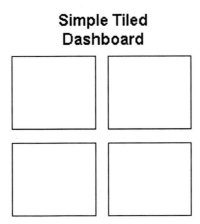

Simple Tiled Dashboard

SAS/GRAPH GREPLAY Procedure Dashboard

In this example, I demonstrate how to use SAS/GRAPH's PROC GREPLAY to create a custom template in order to display 20 graphics together in a very carefully chosen custom dashboard layout. Using the PROC GREPLAY technique described here, you should be able to design custom dashboards with any layout you can imagine.

There are many factors to consider when deciding what graphs to put in your dashboard, and how they should be arranged on the page. That topic is beyond the scope of a simple example like this, and is covered in detail in Stephen Few's excellent book *Information Dashboard Design*.

For this particular dashboard, I designed several graphs to answer the specific questions posed in *DM Review*'s Dashboard contest, and then printed separate copies of each graph (each one on a separate piece of paper). I tried many different physical arrangements, re-arranging the physical pieces of paper on a table top until I had an arrangement I liked. In general, I tried to put the most important graphs toward the top/left, and tried to keep similar graphs side-by-side for easy comparison.

Once I had chosen the general layout, I drew a rough sketch of the layout (by hand, or using a drawing package). Here is the rough sketch:

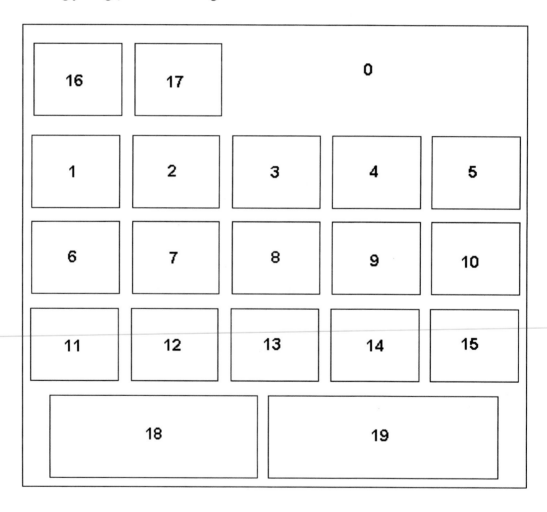

Determining Coordinates

Once you have decided on the general layout, you will need to determine the X and Y coordinates for the four corners of each graph on the page. The units for the X and Y coordinates are in percentages of the screen, from 0 to 100, with the (0,0) being at the bottom/left, and (100,100) at the top/right. I find it helpful to use a printed copy of my rough layout, and physically write the estimated X and Y coordinates on the page, as shown in the image below.

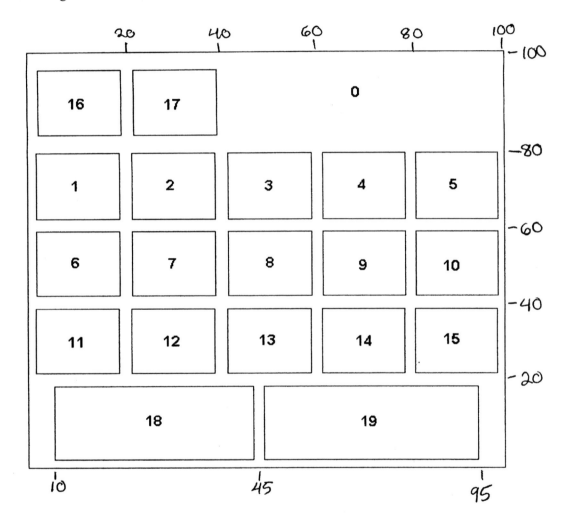

Creating the GREPLAY Template

Once you have the X and Y coordinates for the corners of each graph, you can set up the custom GREPLAY template, by specifying each panel number, followed by the coordinates for the four corners of the graph that goes into that panel.

The number to the left of the "/" is the number ID for the panel in the template (these numbers correspond to the numbered areas in my rough sketch), and the other numbers are the X and Y coordinates for the four corners of the graph (llx=lower left X, lly=lower left Y, ulx=upper left X, uly=upper left Y, and so on). I usually start with nice even numbers for the X and Y values, and then make small adjustments later, if they are needed. GREPLAY templates are very flexible; you can put gaps between the graphs, overlap them, or even specify four corners of a non-rectangular area and let GREPLAY "stretch" the graph out of proportion to fill the area (which is sometimes useful for special effects).

```
proc greplay tc=tempcat nofs igout=work.gseg;
  tdef dashbrd des='Dashboard'

/* overall titles (whole page) */
   0/llx = 0    lly =   0
     ulx = 0    uly =100
     urx =100   ury =100
     lrx =100   lry =   0

/* 2 graphs in top/left */
  16/llx = 0    lly = 80
     ulx = 0    uly =100
     urx =20    ury =100
     lrx =20    lry = 80
  17/llx =20    lly = 80
     ulx =20    uly =100
     urx =40    ury =100
     lrx =40    lry = 80

/* 1st row of 5 graphs */
   1/llx = 0    lly = 60
     ulx = 0    uly = 79
     urx =20    ury = 79
     lrx =20    lry = 60
   2/llx =20    lly = 60
     ulx =20    uly = 79
     urx =40    ury = 79
     lrx =40    lry = 60
{and so on...}
```

During the dashboard development process, I often find it useful to (temporarily) have a visible border outline around each of the rectangular areas. You can easily make the borders visible using the COLOR= option in the GREPLAY template, such as the following:

```
   1/llx = 0    lly = 60
     ulx = 0    uly = 79
     urx =20    ury = 79
     lrx =20    lry = 60
     color=black
```

Now, we can set the size of the dashboard (using GOPTIONS XPIXELS= and YPIXELS=), and calculate the size of the individual graphs by applying the percent values (from the GREPLAY template) to the total size. I chose to make this dashboard 900 pixels wide, and 800 pixels tall:

```
goptions xpixels=900 ypixels=800;
```

To calculate the size of the individual graphs, you apply the percent values (from the GREPLAY template) to the total XPIXELS and YPIXELS values. For example, plot1 extends from 0 to 20% in the GREPLAY template's X-direction, therefore XPIXELS=20% of 900, which is 180. Plot1 extends from 60 to 79% in the GREPLAY template's Y-direction, therefore YPIXELS=(79–60)=19% of 800, which is 152.

```
goptions xpixels=180 ypixels=152;
```

Testing the GREPLAY Template

I often find it useful to "sanity check" my GREPLAY templates by creating a simple or trivial graph, and displaying it into all the areas of the template. This gives me a general idea of what the dashboard will look like. As I finish each of the real graphs, I replace the simple place-holder graphs with the real thing.

As you generate the individual graphs for your dashboard, you will want to save them in GRSEGs, and then use PROC GREPLAY to display each graph (GRSEG) in the desired area of the template. SAS/GRAPH can supply default GRSEG names, but you will want to specify your own (more mnemonic) names for the GRSEGs. That way you will know what the names are so you can use PROCGREPLAY to put them into the template. Specify the name using the NAME= option in each of the SAS/GRAPH procedures.

Here is some code to create a simple graph, and store it in a GRSEG named 'simple':

```
data foo;
input letter $1 value;
datalines;
A 1
B 2
C 3
;
run;

goptions gunit=pct htitle=18 ftitle="albany amt/bold" htext=12
ftext="albany amt";
axis1 label=none value=none major=none minor=none;
axis2 label=none;
pattern1 v=s c=graycc;

title "Simple";
proc gchart data=foo;
vbar letter / type=sum sumvar=value nostats width=15 space=8
 raxis=axis1 maxis=axis2 noframe name="simple";
run;
```

And here is the code to replay the "simple" graph into each of the areas of the GREPLAY template:

```
{some GREPLAY details omitted}
treplay
 16:simple 17: simple 0: simple
 1:simple 2:simple 3:simple 4:simple 5:simple
 6:simple 7:simple 8:simple 9:simple 10:simple
 11:simple 12:simple 13:simple 14:simple 15:simple
 18:simple 19:simple;
```

The results do not look like a great dashboard, but they are useful:

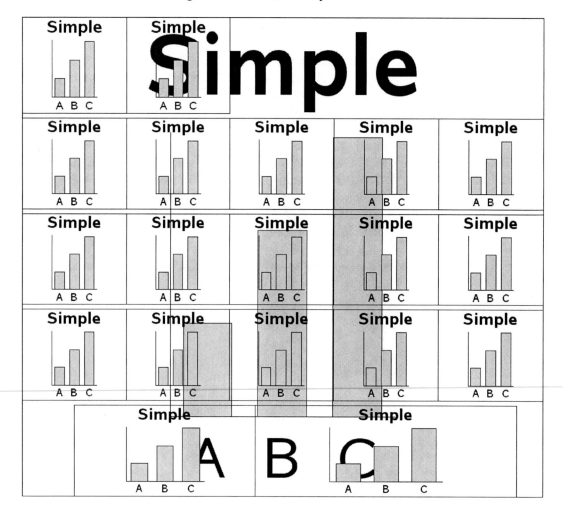

Creating the Individual Graphs

Now we can create the actual graphs. I do not go into all the 1,000-plus lines of code for creating all the individual custom plots (you can download the full sample if you are interested in that), but instead I will show you generally how the title slide and one graph are created. The important thing to know is that you create all the graphs, and store each one in a named GRSEG entry.

The "titles" graph might look simple, but it is actually somewhat complex. PROC GSLIDE will let you easily add titles, but it does not give you much control over the placement of the titles (only left or right/center). Therefore, I use a blank GSLIDE (with no titles), and annotate the title text. This enables me to control the exact size and placement. I also annotate a custom legend on the titles GSLIDE. The titles GSLIDE covers the entire dashboard (to give me total flexibility *if* I want to put something else on the dashboard), but I only wanted the text and legend in the top-right area. If I had wanted text (or lines, logos, and so on) in other locations on the dashboard, I could have also annotated those using this GSLIDE.

Here is the code to generate the Annotate data set, and display it on the GSLIDE:

```
data titlanno;
length function $8 color style $20 text $50;
retain xsys ysys '3' hsys '3' when 'a';
function='label';
position='5';
x=75; y=97; color='black'; size=4; style="&fntb";
text='Sales Dashboard'; output;
{and so on, for the other annotated text and legend}
run;

goptions xpixels=900 ypixels=800;

title;
footnote;
proc gslide anno=titlanno name="titles";
run;
```

The code for plot1-plot17 is basically the same code as was explained in detail in Example 1, "Bar with Annotated Marker." Below is the final code used to generate plot1 (the Market Share plot), and save the results in the plot1 GRSEG entry.

```
goptions xpixels=180 ypixels=160;

title "Market Share";
proc gchart data=metrics anno=targets;
 vbar quarter / discrete type=sum sumvar=a_market_share
 subgroup=ev_market_share  nolegend
 raxis=axis1 maxis=axis2 autoref cref=&crefgray clipref
 coutline=&gray width=14 space=14 name="plot1";
run;
```

One thing you might notice in the code above is that I specify GOPTIONS XPIXELS= and YPIXELS= just before I create each graph. This is important, and affects the proportions of the graph and text. If you do not specify the correct xpixels and ypixels for each graph (in relation to the whole dashboard), then the graph will probably look stretched or squished in the dashboard, and you will possibly have text running outside of the boundaries the graph should be within. If you see strange problems like that, go back and double-check your xpixels and ypixels!

Replaying Graphs into Template

The following code replays the titles and plot1 GRSEG entries into their panels (areas 0 and 1) in the custom dashboard template. As you can see, I am still replaying the simple graph in all the other template panels, and I still have the border color turned on (so the borders are visible). The custom dashboard is starting to take shape.

```
goptions xpixels=900 ypixels=800;

{some greplay details omitted}
treplay
 16:simple 17:simple  0:titles
 1:plot1 2:simple 3:simple 4:simple 5:simple
 6:simple 7:simple 8:simple 9:simple 10:simple
 11:simple 12:simple 13:simple 14:simple 15:simple
 18:simple 19:simple;
```

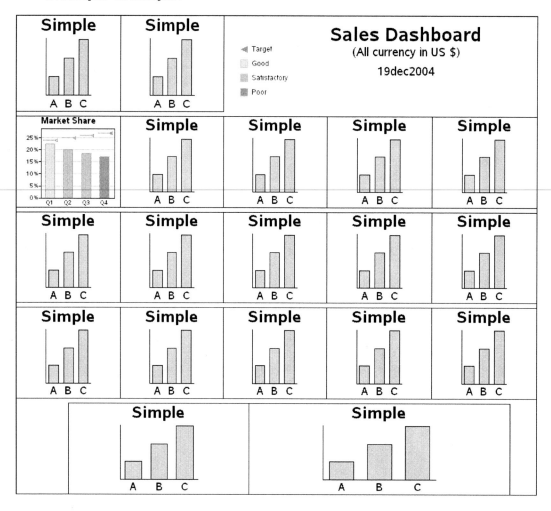

Once you have put all the graphs in their desired positions, you can turn off the border colors around each graph by deleting the COLOR= option from the GREPLAY template code.

Once you have generated all your graphs, and saved them in GRSEG entries, creating the final, complete dashboard is a simple matter of replaying them all into the template, as follows:

```
{some GREPLAY details omitted}
treplay
 16:plot16 17:plot17  0:titles
 1:plot1 2:plot2 3:plot3 4:plot4 5:plot5
 6:plot6 7:plot7 8:plot8 9:plot9 10:plot10
 11:plot11 12:plot12 13:plot13 14:plot14 15:plot15
 18:plot18 19:plot19;
```

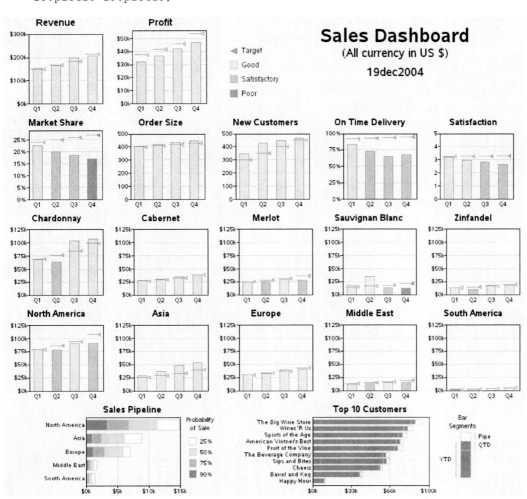

EXAMPLE 3

Paneling Multiple Graphs on a Page

Purpose: Although a simple grid of graphs is not a *dashboard*, it is still a useful way to visualize data in certain situations. This example demonstrates how to tile multiple graphs onto the same page using the ODS HTMLPANEL tagset.

After gasoline prices in the U.S. hit $3 and $4 a gallon in 2008, everyone became interested in finding cars with good mileage. A good Web site I found for looking up the miles per gallon (MPG) is www.fueleconomy.gov. They have a really nice site that provides the MPG data for vehicles sold in the U.S., enabling you to look up the MPG values for an individual vehicle, download a PDF document with the data for all the cars in a given year, or download the data in spreadsheet format.

This example demonstrates how to import their spreadsheet data into SAS and create plots of the MPG data by type of car. As a bonus, I think you will also find the MPG plots very interesting and informative.

Data Preparation

First, get the data, from the http://www.fueleconomy.gov/feg/download.shtml Web page, selecting the "2011 Datafile" link and saving the 11data.zip file to your computer, and then extracting the "2011FEguide-public.xlsx" Excel spreadsheet from the ZIP file. You can then import the data from the Excel spreadsheet into a SAS data set using the SAS/ACCESS to PC Files product to run the following code:

```
proc import datafile="2011FEguide-public.xlsx" dbms=EXCEL out=mpgdata
replace;
 getnames=YES; mixed=YES; scantext=NO;
run;
```

Note: If you do not have SAS/ACCESS to PC Files (and therefore cannot run PROC IMPORT on Excel spreadsheets), an alternative would be to export the data from Excel as a CSV file, and then read that into SAS.

PROC IMPORT uses the column header text as the variable names. You can assign custom text labels to the variables by using label statements in a DATA step, and the labels will be used in the graph axes. You could also do "data cleaning" in the DATA step, if needed, such as the IF statement that eliminates data observations that do not contain the variable CARLINE_CLASS_DESC.

```
data mpgdata; set mpgdata;
 label Hwy_FE__Guide____Conventional_Fu='Highway';
 label City_FE__Guide____Conventional_F='City mpg';
 if Carline_Class_Desc^='' then output;
run;
```

And here is a final bit of data preparation: since we are going to be creating a separate plot for each car classification (that is, plotting by CARLINE_CLASS_DESC), the data set needs to be sorted by that variable.

```
proc sort data=mpgdata out=mpgdata;
by Carline_Class_Desc;
run;
```

Graph the Data

Now that the data is in a SAS data set, we can easily plot it using PROC GPLOT:

```
goptions xpixels=284 ypixels=360;

symbol1 color=blue value=circle h=1.4 interpol=none;

proc gplot data=mpgdata;
 plot
Hwy_FE__Guide____Conventional_Fu*City_FE__Guide____Conventional_F=1;
run;
```

But the results are not much to look at.

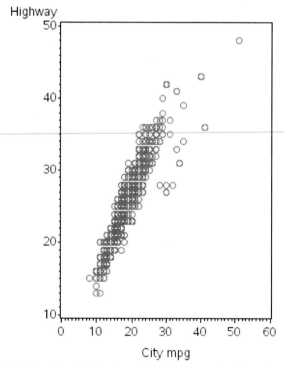

This is a book about custom SAS graphs, therefore let's make a few customizations.

First, let's add a few standard customizations. Rather than letting the axes auto-scale (and pick a slightly different range for the X and Y axes), it is preferable in this case to hardcode the range for the axes so they both show 0–60 MPG. Also, since both axes are showing MPG, and you want to be able to compare them apples to apples, it is good to hardcode a physical length for the axes (LENGTH=2.3in). That way the axes will be drawn proportionally, no matter what the proportions of the PNG file are. The minor tick marks clutter the graph, therefore let's eliminate them by specifying `minor=none` as an axis option. I find it useful to have light gray reference lines at the major tick marks, so I can easily see whether the plot markers are above or below those values; the AUTOHREF and AUTOVREF options add these reference lines.

```
axis1 length=2.3in offset=(0,0) order=(0 to 60 by 10) minor=none;
axis2 length=2.3in offset=(0,0) order=(0 to 60 by 10) minor=none;

proc gplot data=mpgdata;
plot
Hwy_FE__Guide____Conventional_Fu*City_FE__Guide____Conventional_F=1 /
 autohref autovref chref=graycc cvref=graycc
 vaxis=axis1 haxis=axis2;
run;
```

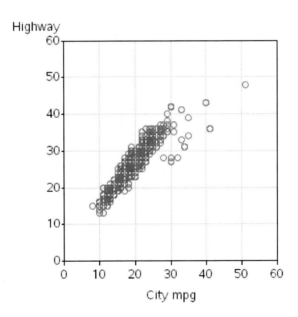

The above plot looks much more refined now, but the plot markers still have no points-of-reference to tell us whether the values are good or bad. This is where custom annotation can be useful.

Annotations

First, let's annotate the text labels "good" in the top-right, and "bad" in the bottom-left. By using the XSYS and YSYS='1' coordinate system, the x and Y coordinates specified are *percent* values, from 0–100 within the axis area.

```
data my_anno;
length function color $8;

 xsys='1'; ysys='1';
 color='gray99';
 position='5'; hsys='3'; when='a'; style=''; size=.;
 function='label'; x=93; y=93; text='good'; output;
 function='label'; x=7; y=7; text='bad'; output;
```

Next, draw a thin (size=.1) gray line from the bottom-left to top-right. If a marker is exactly on this line, the vehicle gets exactly the same MPG in highway driving as they do in city driving. Traditional engines will typically be above this diagonal line, and hybrids will typically be near (or below) the line.

```
 color='graycc';
 when='b';
 function='move'; x=0; y=0; output;
 function='draw'; x=100; y=100; size=.1; output;
```

And finally, draw two wider (size=1) green line segments at 30 MPG—vehicles outside/above this line have "good" MPG, and cars below it are "bad" (note that this is just a totally arbitrary number I picked, based on what I think is good and bad MPG). I use xsys/ysys='2' so that I can specify the x/y values using the same coordinate system as the data (a 30 here is 30 MPG). Note that it is especially important for this line to use when='b' so that it is drawn before/behind everything else. Otherwise, this thick line would obscure plot markers.

```
 xsys='2'; ysys='2';
 color='cx00ff00';
 function='move'; x=30; y=0; output;
 function='draw'; x=30; y=30; size=1; output;
 function='draw'; x=0; y=30; size=1; output;
run;
```

Here is what the annotated text and graphics look like, plotted on a blank set of axes:

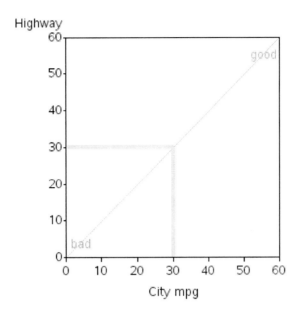

Simply specify the ANNO= option when you run PROC GPLOT to have the annotation show up on the graph with the plot markers:

```
proc gplot data=mpgdata anno=my_anno;
   {same code as before}
```

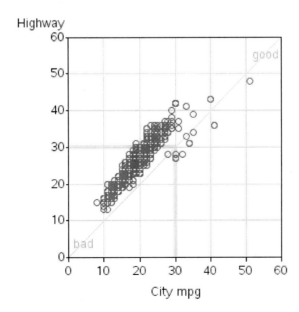

One remaining weakness with the graph is that it is trying to plot "too much" in one plot.

There are almost 1000 vehicles in the data, and many of the markers are overprinting other markers in exactly the same locations. Also, most people shopping for a car would be interested in seeing a plot of all the cars in a certain classification (small, medium, large, and so on).

Since the data is sorted by CARLINE_CLASS_DESC, we can easily plot it by that variable. One enhancement I like to make is that instead of using the default by-titles (such as `Carline_Class_Desc='Compact Cars'`), I suppress the automatic titles by specifying `options nobyline`, and I use the by-value in my own title statement.

```
options nobyline;
title1 ls=1.5 "#byval(Carline_Class_Desc)";

proc gplot data=mpgdata anno=my_anno;
by Carline_Class_Desc;
plot
Hwy_FE__Guide____Conventional_Fu*City_FE__Guide____Conventional_F=1 /
 autohref autovref chref=graycc cvref=graycc
 vaxis=axis1 haxis=axis2;
run;
```

You now get a plot for each car classification, arranged sequentially (one after the other) down the Web page. The plot markers are much less crowded and easier to read, but you would have to do a lot of scrolling up and down in your Web browser to see all 19 plots. Here are two of those plots:

Compact Cars

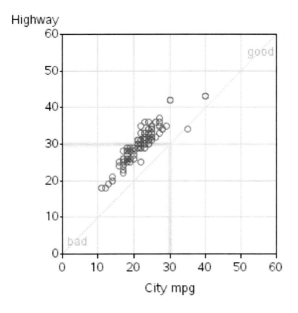

Large Cars

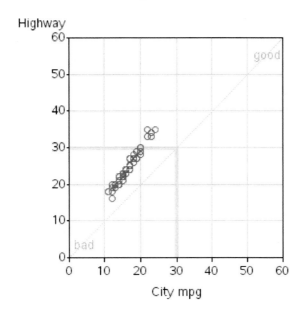

And so on, for all the other 17 car classifications.

Paneling the Graphs

Wouldn't it be nice to put the plots in a panel three-across? That way you could see and compare more plots at the same time, rather than scrolling up/down to see them. This can be done with just a little more work, using the ODS HTMLPANEL tagset. Here is the basic code to display the graphs paneled three-across:

```
ODS LISTING CLOSE;
%let panelcolumns=3;
ods tagsets.htmlpanel path="." (url=none) file="somename.htm"
style=minimal;

{do your graphs}

ods tagsets.htmlpanel event = panel(finish);
quit;
ods html close;
```

Now I am going to make it just a little fancier, by creating a title using PROC GSLIDE to go above all the graphs. First I will show you just the skeleton (without any GSLIDE or GPLOT details, so you can easily see how it fits into the ODS HTMLPANEL code). And then I will show you the full-blown code:

```
ODS LISTING CLOSE;
%let panelcolumns=3;
ods tagsets.htmlpanel path="." (url=none) file="somename.htm"
style=minimal;

goptions xpixels=830 ypixels=80;
{do your title gslide}

ods tagsets.htmlpanel event = panel(start);
goptions xpixels=284 ypixels=360;
{do your graphs}

ods tagsets.htmlpanel event = panel(finish);
quit;
ods html close;
```

And here is the full-blown code, with all the details:

```
%let year=2011;
%let name=gas&year;

ODS LISTING CLOSE;

%let panelcolumns=3;
ods tagsets.htmlpanel path="." (url=none) file="&name..htm"
(title="&year Gas Mileage Plots") style=minimal;

goptions device=png;
goptions ftitle="albany amt/bold" ftext="albany amt" htitle=11pt
htext=10pt;

goptions xpixels=830 ypixels=80;
title1 h=15pt "Gas Mileage Plots";
title2 h=12pt ls=1.5 font="albany amt/bold" "year &year";
proc gslide name="&name";
run;
```

```
ods tagsets.htmlpanel event = panel(start);
goptions xpixels=284 ypixels=360;
options nobyline;


axis1 length=2.3in offset=(0,0) order=(0 to 60 by 10) minor=none;
axis2 length=2.3in offset=(0,0) order=(0 to 60 by 10) minor=none;


symbol1 color=blue value=circle h=1.4 interpol=none;


title1 ls=1.5 "#byval(Carline_Class_Desc)";
footnote h=3pct " ";


proc gplot data=mpgdata anno=my_anno;
by Carline_Class_Desc;
plot
Hwy_FE__Guide____Conventional_Fu*City_FE__Guide____Conventional_F=1 /
 autohref autovref chref=graycc cvref=graycc
 vaxis=axis1 haxis=axis2
 des="" name="gas_mpg_&year._#byval(Carline_Class_Desc)";
run;


ods tagsets.htmlpanel event = panel(finish);


quit;
ods html close;
```

This code produces the following output (note that this is just a screen-capture of the first two rows, for brevity—there are seven total rows of plots in the final output):

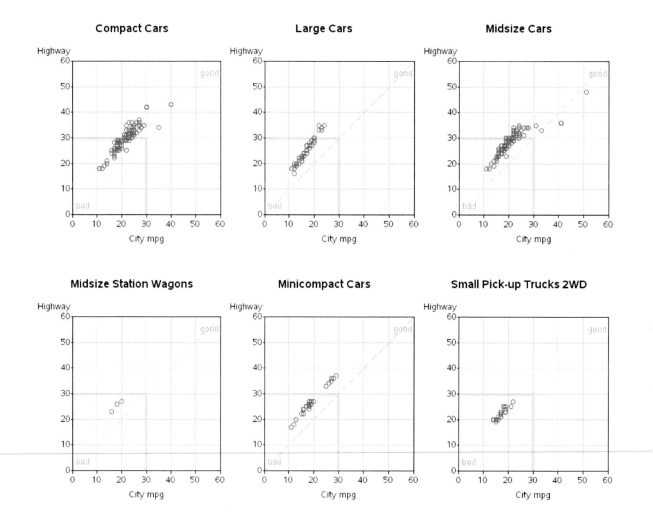

EXAMPLE 4

Using an Image As Part of the Graph

Purpose: Demonstrate how to use an image as part of the graph itself.

Although a simple graph is usually best for analyzing data, sometimes a visually captivating graph is needed for certain situations or audiences. Sometimes you need a graph that will catch people's attention and lure them into reading an article, or stopping on a webpage, to learn more about the data shown in the captivating graph.

When you need to create such a graph, I encourage you to try to add *fanciness* that is at least related to the data, rather than simply using random graphical tricks like bright colors, glittery images, three-dimensionality, or what is sometimes called "dancing bologna."

The graph described here is a good example. I first saw a graph similar to this one on the (now-defunct) uppervalleybullion.com Web site. This trick graph not only caught my attention, but it also immediately let me know it had something to do with the shrinking value of the dollar.

As with the other examples in this book, there is no simple procedure option that can be used to generate this type of graph. But by using annotate, and a tiny bit of custom programming, this visually captivating graph is possible.

First, I located some consumer price index data on the Bureau of Labor Statistics (BLS) Web site that could be used to represent the value of a dollar over time. With about 100 years of data, and 12 monthly values for each year, that is over 1000 lines of data, so I show only a subset of the data below (but you can get the whole series from this author's Web site or the BLS Web site)[1]:

```
data mydata;
input year month cpi;
datalines;
1913       1        9.8
1920       1       19.3
1930       1       17.1
1940       1       13.9
1950       1       23.5
1960       1       29.3
1970       1       37.8
1980       1       77.8
1990       1      127.4
2000       1      168.8
2010       1      216.687
;
run;
```

Next I calculate some macro variables. I determine the minimum year (MINYR), and then the minimum consumer price index (COMPCPI) from that year. This will be the CPI that I compare to all the other yearly values. There are several alternative ways to do this, and I chose to do it step-by-step using PROC SQL.

```
proc sql;
select min(year) into :minyr from mydata;
select min(cpi) into :compcpi from mydata where year=&minyr;
quit; run;
```

I trim the blanks off the MINYR, so that it will fit nicely into the title string later. Alternatively, I could have used `separated by ' '` in the sql query to get rid of the blanks when the macro variable was created. Both techniques are useful.

```
%let minyr=%trim(&minyr);
```

I need to combine the year and month into a single value in order to plot the data the way I want. I could have built a string containing the year and month and read it in as a SAS date, but in this case I decided to use a more direct approach and calculate the YEAR + FRACTION_OF_YEAR (which I call YEARPLUS). While running the data through this DATA step, I also go ahead and calculate the dollar's value each year. After I am done, I stuff the minimum and maximum YEARPLUS values into macro variables (called MINYEAR and MAXYEAR) so I can easily use them later.

```
data mydata; set mydata;
 yearplus=year+((month-1)/12);
 dollarvalue=&compcpi/cpi;
run;
proc sql;
select min(yearplus) into :minyear from mydata;
select max(yearplus) into :maxyear from mydata;
quit; run;
```

At this point the data is ready to plot.

```
symbol1 v=none i=join c=black;

title1 ls=2.5 "Value of a U.S. Dollar";
title2 font="albany amt/bold" "Compared with a &minyr dollar";

proc gplot data=mydata;
 plot dollarvalue*yearplus;
run;
```

Value of a U.S. Dollar
Compared with a 1913 dollar

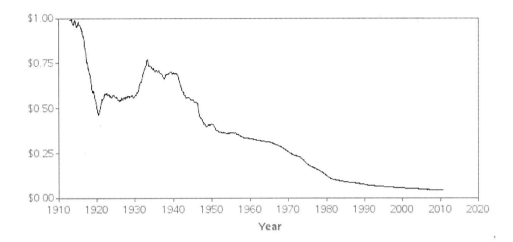

But the basic line plot can be improved. PROC GPLOT lets you color the area below the line using the AREAS= option, but it does not let you specify an image for the fill pattern of the area under the curve. And using a color, as shown below, just does not have the same impact as using the image of a dollar, as was used in the original graph I am imitating.

```
symbol1 v=none i=join c=black;
pattern1 v=solid c=green;
proc gplot data=mydata;
 plot dollarvalue*yearplus / areas=1;
run;
```

Value of a U.S. Dollar
Compared with a 1913 dollar

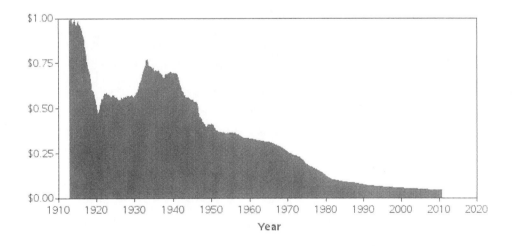

To get the desired graph, we will need to do some custom programming. You can annotate an image anywhere on the graph, so why not annotate an image of a dollar in the exact same area where the data is plotted. Note that it would be easier to use the built-in IFRAME= GPLOT option, but then the dollar would fill the entire axis area, and not necessarily limit itself to the area occupied by the dollar data. Therefore, you must use annotate to guarantee the desired effect.

To annotate an image, you first move to the bottom-left x and coordinates, and then specify the IMAGE function and the top-right coordinates. The image then fills this rectangular area. In the x direction I specify `xsys='2'` so that I can use the minimum and maximum YEARPLUS values, and so that the dollar image does not extend beyond the data range. In the y direction I specify `ysys='1'` so I can easily specify values to make the image extend from the bottom to the top of the axes (y=0 to y=100). And I specify `when='b'` so the annotated dollar is drawn first and appears behind the rest of the graph.

```
data annodollar;
length function $8;
xsys='2'; ysys='1'; when='b';
function='move';  x=&minyear; y=0; output;
function='image'; x=&maxyear; imgpath='dollar_image.jpg'; style='fit';
y=100; output;
run;

symbol1 v=none i=join c=black;
proc gplot data=mydata anno=annodollar;
 plot dollarvalue*yearplus;
run;
```

Value of a U.S. Dollar
Compared with a 1913 dollar

Now, the only thing left to do is to cover (obscure) the part of the dollar that is above the plot line. To do this, we return to the built-in GPLOT AREAS= option. This time we tell it `areas=2` so that both the area below *and* above the one graph line are filled with a pattern or color. But we get a little tricky with the patterns: we specify that the area below the line is filled with an empty pattern (in other words, you can see through it), and the area above the line is filled with solid white color, same as the background.

```
goptions cback=white;

symbol1 v=none i=join c=black;

pattern1 v=empty c=pink;
pattern2 v=solid c=white;

axis1 label=none order=(0 to 1 by .25) minor=none offset=(0,0)
length=2.5in;
axis2 label=(font="albany amt/bold" 'Year') minor=none offset=(0,0)
length=6.0in;

proc gplot data=mydata anno=annodollar;
  format dollarvalue dollar5.2;
  plot dollarvalue*yearplus / areas=2
   vaxis=axis1 haxis=axis2;
run;
```

Value of a U.S. Dollar
Compared with a 1913 dollar

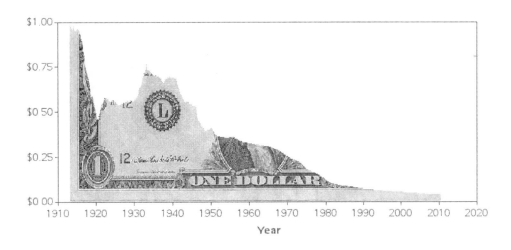

Notes

[1] Bureau of Labor Statistics (BLS) Website: ftp://ftp.bls.gov/pub/special.requests/cpi/cpiai.txt.

EXAMPLE 5

Customizing a Map

Purpose: Describe how to combine areas of a map, and label areas on the map.

Two topics that frequently come up are customizing map borders and labeling map areas. This example provides some very useful (and re-usable) techniques for both of those tasks.

Maps of most countries are included in your SAS/GRAPH installation. There is generally only one map per country, and it shows the most granular (smallest) areas available for each country. For example, maps.germany has about 400 areas, if you plot the map as is:

```
proc gmap map=maps.germany data=maps.germany;
 id id;
 choro id / levels=1 nolegend coutline=gray55;
run;
```

Maps.Germany

But what if you do not want a map at the most-granular level? What if, for example, you want a map of Germany showing the 16 states? Not to worry—SAS/GRAPH can do that.

Most of the SAS/GRAPH maps have a corresponding feature table (usually named the same as the map, with a 2 appended to the end of the name). The feature tables contain additional information, such as the full-text names to go with the ID numbers, and sometimes other (higher-level) geographical groupings that each ID number is a member of. For maps.germany, the feature table is named maps.germany2, and it contains the state name (STATE2) for each ID number.

Therefore, we can easily generate a map of Germany colored by the state names, as follows:

```
legend1 position=(right middle) across=1 label=(position=top j=c)
shape=bar(.15in,.15in);

proc gmap map=maps.germany data=maps.germany2;
 id id;
 choro state2 / discrete coutline=gray legend=legend1;
run;
```

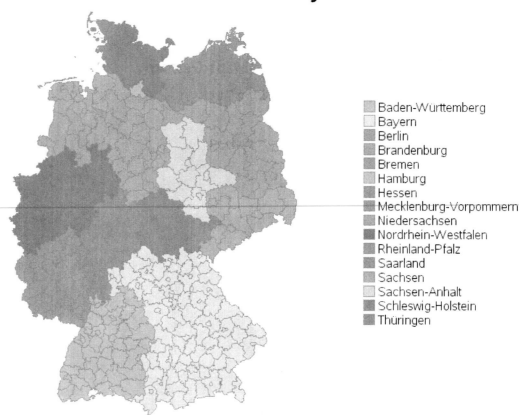

Although the map is shaded by the state name, you probably notice all the borders of the areas within the states. With a little custom programming, you can get rid of these internal borders using PROC GREMOVE.

First, you must add the state names to the map data set, and then sort by the state name (you need it to be sorted, so that you can sort by STATE2 later). Although you're sorting by STATE2, we want the observations within each STATE2 to remain in the order they were in originally (so the points along the border of the states are connected in the right order). Therefore, I first create a variable to remember the ORIGINAL_ORDER so I can re-sort the observations within each state back in their original order.

```
data germany; set maps.germany;
 original_order=_n_;
run;

proc sql;
 create table germany as
 select unique germany.*, germany2.state2
 from germany left join maps.germany2
 on germany.id=germany2.id
 order by state2, original_order;
quit; run;
```

Now you can use PROC GREMOVE to remove the internal borders within each state. The BY statement indicates the variable that uniquely identifies the areas of the new map, and the ID statement indicates the variable that uniquely identifies the areas of the original map,

```
proc gremove data=germany out=germany;
 by state2;
 id id;
run;
```

And you can plot the new map using almost the same code as before, but now using STATE2 as the ID variable:

```
proc gmap map=germany data=germany;
 id state2;
 choro state2 / discrete coutline=gray legend=legend1;
run;
```

16 States of Germany

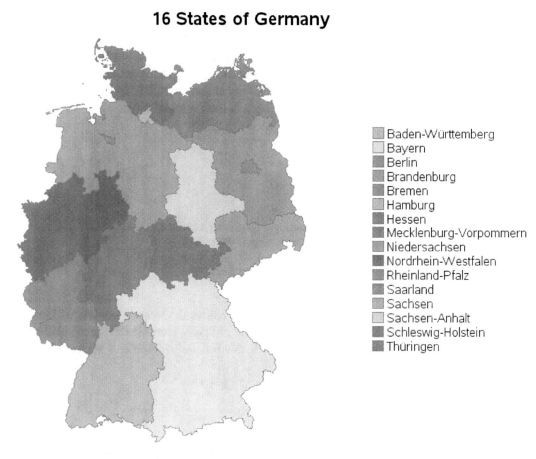

Adding labels is fairly simple to do using the Annotate facility. Basically just determine the X and Y map coordinates where you want the label, and then annotate the desired text in that location using the LABEL function.

There are various ways to estimate the X and Y coordinates at the center of each map area. For this example, I use the %CENTROID macro and store the centroid coordinates in the ANNO_SIMPLE data set. I then convert the ANNO_SIMPLE data set into an Annotate data set by telling it the X and Y coordinates are in the same coordinate system as the map borders (xsys/ysys=2), and telling it to use the state name (STATE2) as the text. This requires very few lines of code, but you must know which lines of code to use.

```
%annomac;
%centroid(germany, anno_simple, state2);

data anno_simple; set anno_simple;
 xsys='2'; ysys='2'; hsys='3'; when='a'; size=2;
 function='label'; text=trim(left(state2));
run;

proc gmap map=germany data=germany anno=anno_simple;
 id state2;
 choro state2 / discrete coutline=gray legend=legend1;
run;
```

That was fairly simple, but unfortunately the map centroids are not always in the best location for the map labels, and sometimes the labels collide. For example, the labels for Berlin and Brandenburg overlap.

16 States of Germany

So, let's try it one more time.

This time, when we build the Annotate data set, we'll hard-code some offsets for certain labels. In a simple situation, you might use the annotate position function to easily adjust the placement of the text (it provides 15 pre-defined adjustments), but using the X and Y offset technique described below gives you <u>more</u> control, and allows you to place the text in the exact desired position.

Here's the algorithm:

- move to the X and Y centroid

- add some X and Y offsets for certain labels using the percent-based coordinate system (xsys/ysys=b)

- move to that new location

- move your text cursor to that new location (CNTL2TXT)

- output the label

- view the output and adjust the X and Y offsets as needed

```
%annomac;
%centroid(germany, anno_names, state2);

data anno_names; set anno_names;
 length function $8;
 hsys='3'; when='a'; size=2;
 xsys='2'; ysys='2'; function='move'; output;
 x=0; y=0;
 text=trim(left(state2));
 /* Apply offsets for certain labels */
 if text='Brandenburg' then y=-1.5;
 if text='Niedersachsen' then y=y+1.5;
 if text='Sachsen' then x=-1;
 xsys='b'; ysys='b'; function='move'; output;
 function='cntl2txt'; output;
 function='label'; output;
run;

proc gmap map=germany data=germany anno=anno_names;
 id state2;
 choro state2 / discrete coutline=gray legend=legend1;
run;
```

And now you now have the following wonderful map, with all the labels in their optimum location:

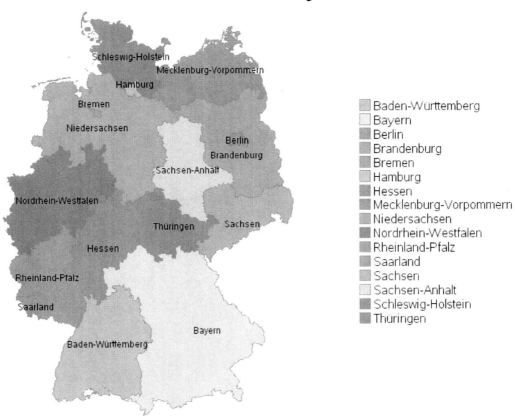

EXAMPLE 6

Overlay G3D Surfaces

Purpose: Overlay multiple 3D surfaces in the same plot.

One of the limitations of the SAS/GRAPH PROC G3D surface plot is that it allows you to plot only a single surface. A colleague in grad school had a need to plot two surfaces on the same set of axes, and I devised a way to do that using PROC GREPLAY. The technique is fairly straightforward, but there are a few caveats.

The data is not really important (you'll have your own data); therefore, I use the following data, which I borrowed from a SAS/GRAPH example:

```
data raw_data;
input X Y Z;
datalines;
-1.0 -1.0   15.5
 -.5 -1.0   18.6
  .0 -1.0   19.6
  .5 -1.0   18.5
 1.0 -1.0   15.8
-1.0  -.5   10.9
 -.5  -.5   14.8
  .0  -.5   16.5
  .5  -.5   14.9
 1.0  -.5   10.9
-1.0   .0    9.6
 -.5   .0   14.0
  .0   .0   15.7
  .5   .0   13.9
 1.0   .0    9.5
-1.0   .5   11.2
 -.5   .5   14.8
  .0   .5   16.5
  .5   .5   14.9
 1.0   .5   11.1
-1.0  1.0   15.8
 -.5  1.0   18.6
  .0  1.0   19.5
  .5  1.0   18.5
 1.0  1.0   15.8
;
run;
```

The raw data contains only 25 data points. Therefore, PROC G3D would produce only the following low-resolution surface (with 25 vertices), which is somewhat lackluster.

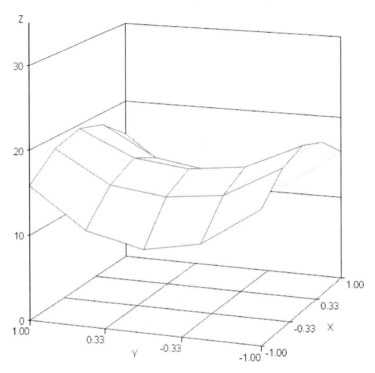

G3D Surface (raw data)

To give PROC G3D more points to work with (and produce a smoother higher-resolution plot), you will want to run the data through PROC G3GRID. This uses a smoothing algorithm, and interpolates extra data points along the smoothed surface. The following PROC G3GRID settings, for example, produce a data set containing 441 "smoothed" data points along the surface.

```
proc g3grid data=raw_data out=smoothed;
grid y*x=z / spline smooth=.05
 axis1=-1 to 1 by .1
 axis2=-1 to 1 by .1;
run;
```

If you were now to plot the smoothed data, the resulting surface plot would look much nicer:

G3D Surface (smoothed data)

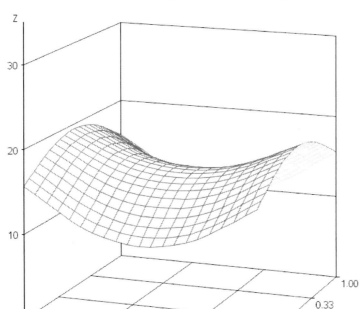

For this example, I need data for two surfaces. To keep the example simple, I am just reusing the same data for both surfaces (applying a +/-8 offset in the Z direction). Of course, you will be using your real data, rather than using this trick to create fake data for the second surface.

```
data smoothed; set smoothed;
 label z1='Z' z2='Z';
 z1=z-8;
 z2=z+8;
run;
```

Now I am ready to start creating the real plots. I create a separate plot for each surface, and later use PROC GREPLAY to display both surfaces together.

You will want everything in both plots to be in the exact same locations, except for the 3-D surfaces (this includes both the text and the axes). Also, as you are creating these individual plots, turn off the displaying of graphs (using the NODISPLAY GOPTION) so they will not be written out until you do the final GREPLAY.

In addition to having all the text and axes in exactly the same positions in both plots, here is a little caveat that most people do not know: if you overlay two plots that both have anti-aliased black text (that is, smoothed edges), then the anti-aliasing is combined and produces a "fuzzy" halo-effect around the text. To avoid this problem, you can make the text on the first plot white (same color as the background). You might want to make the text light gray while you are experimenting with the code and getting everything lined up just right (as shown below), but when you are ready to create the final graph, change it to white. Here is the code for the first plot (PLOT1).

```
goptions nodisplay;
goptions ctext=white;

title ls=1.5 "Overlay Multiple G3D Surfaces, using GReplay";
proc g3d data=smoothed;
 plot y*x=z1 /
 grid zmin=0 zmax=30 xticknum=4 tilt=80
 ctop=purple cbottom=cx00ff00 name="plot1";
run;
```

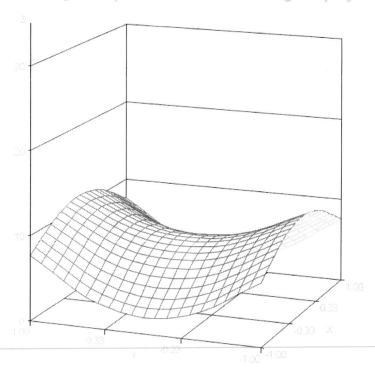

The code for the second plot (PLOT2) is very similar, but with black text:

```
goptions ctext=black;

proc g3d data=smoothed;
 plot y*x=z2 /
 grid  zmin=0 zmax=30 xticknum=4 tilt=80
 ctop=blue cbottom=red name="plot2";

run;
```

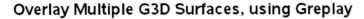

Overlay Multiple G3D Surfaces, using Greplay

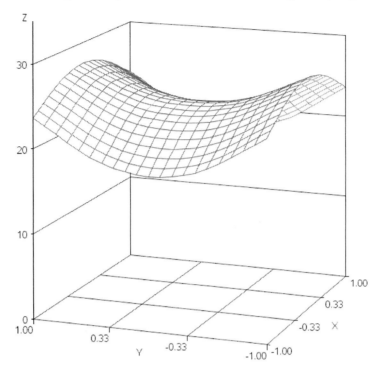

Now that you have created both plots, and stored them as named GRSEG entries (PLOT1 and PLOT2), you can overlay the two graphs on the same page using PROC GREPLAY. The following code creates a GREPLAY template with one area that covers the entire page (0,0 to 100,100). The area is called area number 1, and I TREPLAY PLOT1 and PLOT2 into each into area (1:plot1 1:plot2).

```
goptions display;

proc greplay tc=tempcat nofs igout=work.gseg;
tdef WHOLE des="my template"
        1/llx=0    lly=0
          ulx=0    uly=100
          urx=100 ury=100
          lrx=100  lry=0
          ;
template = whole;
treplay 1:plot1 1:plot2;
run;
```

Overlay Multiple G3D Surfaces, using Greplay

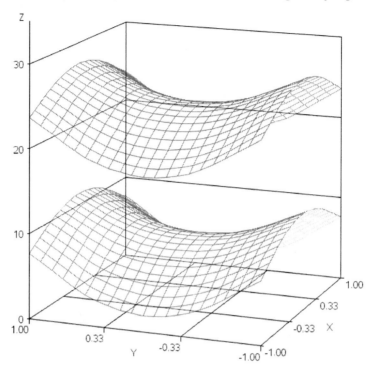

One issue to keep in mind: the G3D surfaces are generated as solid surfaces (blocking out everything behind them), but when you overlay them using GReplay, one copy of the axes is printed on top of the surface. This makes the surface look transparent when it comes to the axis, but not transparent when it comes to the surface crossing over itself. Most people will not notice this problem, but it is a limitation you should be aware of. Also, this trick is probably viable only for plotting two separate (non-overlapping and non-intersecting) surfaces.

EXAMPLE 7

Adding Border Space

Purpose: Describe tricks that can be used to add the desired amount of white space on all sides of a graph.

Of all the SAS/GRAPH tricks that I use, this is probably the one I use most frequently.

It is often very useful to be able to control the amount of white space on the top, bottom, left, and right of a graph. One way to do that is to use blank TITLE and FOOTNOTE statements. In this example, I use a very simple bar chart to demonstrate the technique, but do not let the simplicity of the bar chart over-shadow the **power** of the technique I will be using to control the white space.

Here is the code for the simple default bar chart:

```
axis1 label=none minor=none;
axis2 label=none value=(angle=90);

pattern1 v=solid color=grayee;

proc gchart data=sashelp.class;
vbar name / type=sum sumvar=height descending
 raxis=axis1 maxis=axis2;
run;
```

By default, SAS/GRAPH output tends to try to fill all the available space. But sometimes this does not produce a "visually appealing" graph — it looks a bit crowded with so little white space around the borders.

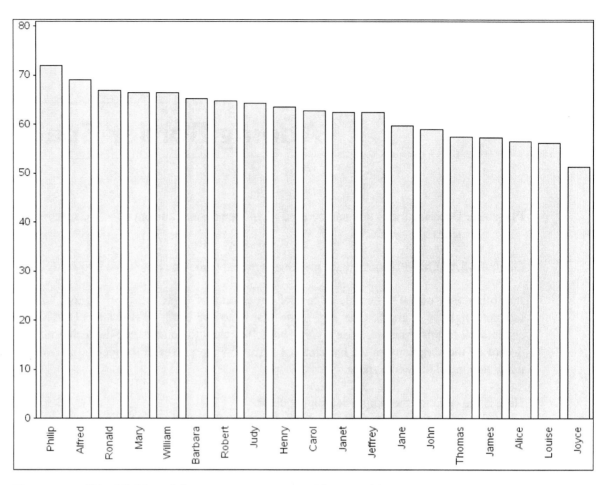

You can use "blank" title and footnote statements to add extra white space.

When using this technique, I typically specify that `goptions gunit=pct`, and specify the height of these blank titles as a percent of the page. You can also specify the height in points. I prefer percent, however, because that makes it easier to change the size of the graph without having to modify the code. In addition to the text height, you can also specify a large LS= value (line spacing) on the title. For now, I am using visible text in the titles, to show you the height value I used and let you see how much space that text occupies. (At the end I will also show the graph with blanked-out text.)

```
goptions gunit=pct ftitle="albany amt/bold" ftext="albany amt";

title1 ls=3.5 "Main Title, with ls=3.5";
title2 h=5 "title2 h=5";

footnote1 h=10 "footnote1 h=10";
footnote2 h=2 "footnote2 h=2";
```

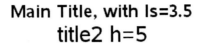

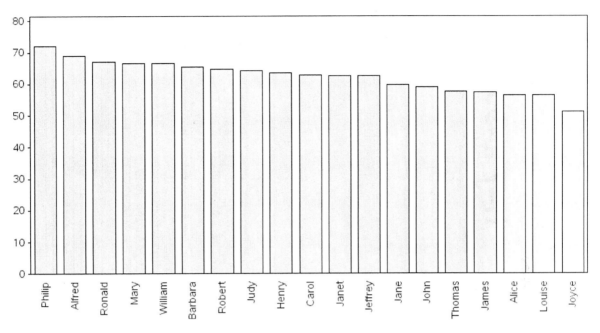

Did you know that you can specify `angle=90` for the title, and the title text is positioned along the left side of the graph? And, of course, if that title is blank, it will add white space to the left of the chart.

```
title1 h=7 a=90 "h=7 a=90";
footnote;
```

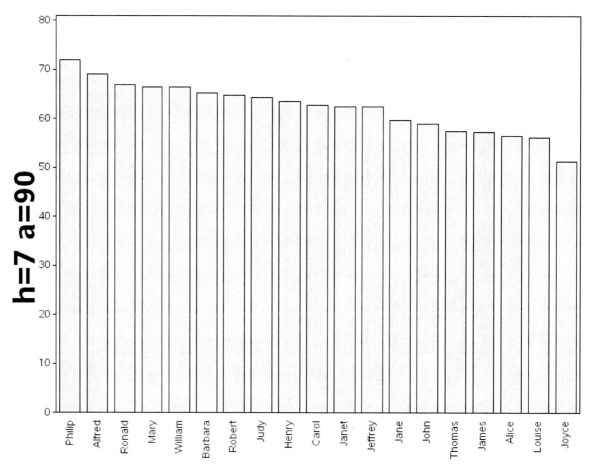

Similarly, if you specify angle=-90, it will position the title on the right side of the graph.

```
title1 h=15 a=-90 "h=15 a=-90";
footnote;
```

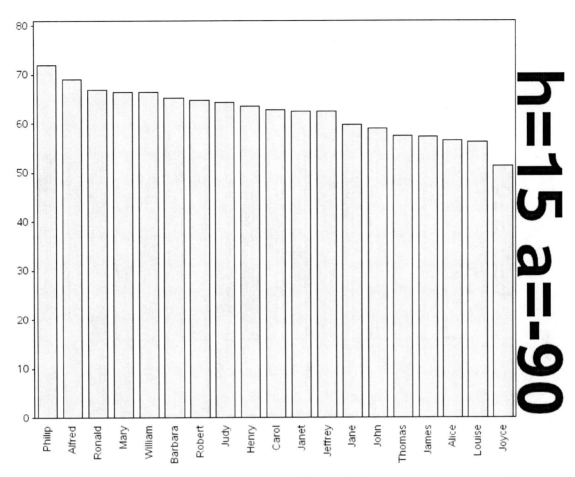

Now, here is the graph with *all* of the previous titles and footnotes, so you can see the combined effect:

```
title1 ls=3.5 "Main Title, with ls=3.5";
title2 h=5 "title2 h=5";
title3 h=7 a=90 "h=7 a=90";
title4 h=15 a=-90 "h=15 a=-90";

footnote1 h=10 "footnote1 h=10";
footnote2 h=2 "footnote2 h=2";
```

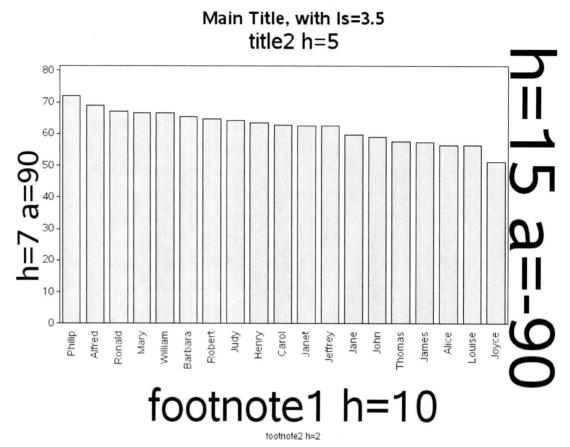

And if you make the text blank, here is the white space added by all those blank titles and footnotes. This is a little more white space than you would normally want, but I am adding extra-large spaces to exaggerate the effect and make it easier to see.

```
title1 ls=3.5 "Same Graph, with Blank Titles";
title2 h=5 " ";
title3 h=7 a=90 " ";
title4 h=15 a=-90 " ";

footnote1 h=10 " ";
footnote2 h=2 " ";
```

Same Graph, with Blank Titles

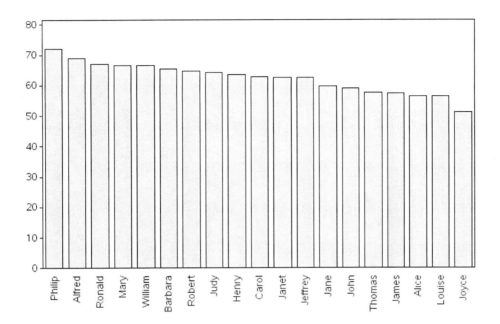

Let me repeat: of all the SAS/GRAPH tricks that I use, this is probably the one I use most frequently.

EXAMPLE 8

Determine Customers Within Map Borders

Purpose: Demonstrate how to use the new SAS/GRAPH PROC GINSIDE to determine whether customers are in a sales area composed of a group of counties.

To make this example easy for you to run, and so you do not have to worry about having a customer database with hundreds of customer locations, I create some fake customer data from the sashelp.zipcode data set that is shipped with SAS:

```
proc sql;
  create table customers as
  select unique city, statecode as state, zip
  from sashelp.zipcode
  where statecode='NC';
quit; run;
```

I then use PROC GEOCODE (new in SAS 9.2) and look up the latitude and longitude of the ZIP code centroid for each customer. If you have the street address of your customer, you could also geocode down to the street level, but you will need to have SAS 9.2 third maintenance release or higher, and you will need to download the lookup table.

Note: The GEOCODE procedure's ZIP option will be changing to METHOD=ZIP in future releases.

```
proc geocode ZIP NOCITY data=customers out=customers;
run;
```

PROC GEOCODE returns the longitude and latitude (X and Y) in degrees, and you need to convert them to radians so they can be conveniently combined with the maps supplied by SAS and projected with PROC GPROJECT. Note that I also multiply the longitude by –1. This converts it to the same eastlong and westlong direction as the maps supplied by SAS. (Alternatively, you could convert the map's longitude and latitude values from radians to degrees, and multiply the map's longitude by –1, and specify the DEGREES option in PROC GPROJECT. However, that seems like a lot more work.)

```
data customers; set customers;
 x=(atan(1)/45 * x)*-1;
 y=(atan(1)/45 * y);
run;
```

The data set will have several extra variables (from PROC GPROJECT), but the important part is that you now have an X and Y location (in radians) for each customer:

```
ZIP     X       Y
27513 1.37528 0.62492
27013 1.40808 0.62387
```

The following code converts the data set into Annotate instructions that will create a dark gray X on the map to represent each customer. We do not use the annotation just yet, but this is a convenient time to create it.

```
data customers;
  set customers;
    length text function $8 style color $20;
    retain xsys ysys '2' hsys '3' when 'a';
  anno_flag=1;
  function='label';
  style='"albany amt"';
  position='5';
  size=2.0;
  color='gray55';
  text='X';
  output;
run;
```

Now to create the map. We start with the US county map (maps.counties). It contains the county outlines for each of the 50 states, and we will work with just the state of North Carolina (the location of the SAS headquarters), and identify a sales region that comprises several counties. Below is the state map with the counties of the sales region colored in red. (Note that for this example you must use maps.counties rather than maps.uscounty because only maps.counties has the unprojected latitude and longitude.)

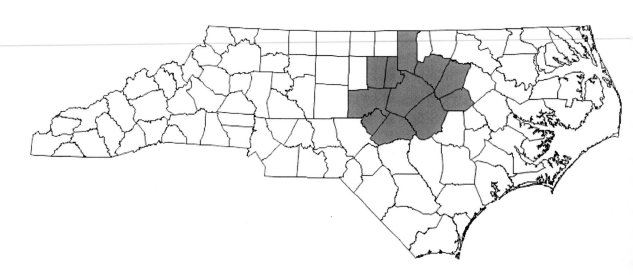

First, select just the counties in the sales region. In the US county map, the counties are identified by their numeric state and county FIPS codes. (If you are uncertain what the numeric FIPS codes are for your counties, you can look them up using their plain-text county names in the maps.cntyname data set.) Then assign as a common variable (REGION_ID) the same value for all the counties in your sales region. In this example I used region_id=999999.

```
data sales_region; set maps.counties (rename=(state=num_state)
  where=(num_state=stfips('NC') and county in (183 101 85 105 37 63
135 77 69 127 195)));
 state=fipstate(num_state);
 region_id=999999;
run;
```

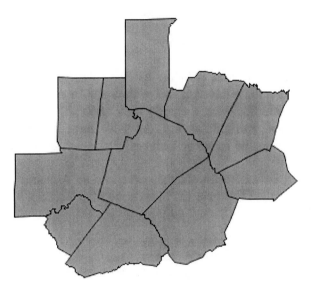

Now convert this group of counties into one contiguous map area by removing all the internal boundaries, using PROC GREMOVE. The ID statement tells GREMOVE what variables are needed to uniquely identify the areas in the original map (ID STATE COUNTY) and the BY statement tells GREMOVE which map areas to group together (all areas with the same REGION_ID).

```
proc gremove data=sales_region out=sales_region;
 by region_id;
 id state county;
run;
```

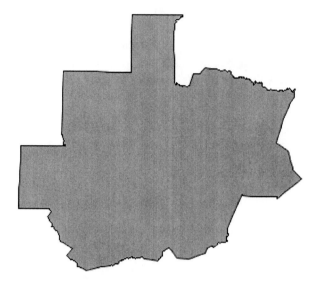

We now have a map area with the external borders of the sales region, and we could hypothetically combine this map area with the NC county map. But overlapping map areas can produce problems. What we really want is the NC county map, with just the *outline* of the sales region overlaid on top. As with most "overlay" challenges, this can be accomplished using the Annotate facility.

The following code converts the map area into an Annotate data set using the technique described in detail in Example 12, "Annotated Map Borders."

```
data sales_region;
   length COLOR FUNCTION $ 8;
   retain XSYS YSYS '2' COLOR 'red' SIZE 1.75 WHEN 'A' FX FY FUNCTION;
   set sales_region; by region_id Segment;
   anno_flag=2;
   if first.Segment then do;
      FUNCTION = 'Move'; FX = X; FY = Y; end;
   else if FUNCTION ^= ' ' then do;
      if X = .  then do;
         X = FX; Y = FY; output; FUNCTION = ' '; end;
      else FUNCTION = 'Draw';
   end;
   if FUNCTION ^= ' ' then do;
      output;
      if last.Segment then do;
         X = FX; Y = FY; output; end;
   end;
run;
```

The resulting Annotate data set (SALES_REGION) produces the following polygon:

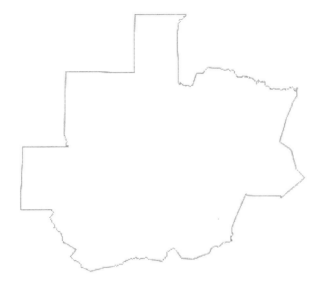

The SALES_REGION data set is not only a valid Annotate data set for drawing the sales region polygon; technically, it is also a valid SAS/GRAPH map data set. Although we are not going to use it to draw a map in PROC GMAP, we *will* specify it using MAP= for PROC GINSIDE.

PROC GINSIDE takes a map and some point data and tells you which map area the points are in. In this case, we use SALES_REGION as the map, and the CUSTOMERS data as the points. Since we have only one sales region, PROC GINSIDE basically tells us whether each customer is inside or outside of that sales region.

Note: PROC GINSIDE was new in SAS 9.2.

```
proc ginside data=customers map=sales_region out=customers;
  id region_id;
run;
```

Here is what a tiny bit of the data would look like before running PROC GINSIDE (showing only two observations from the data set):

```
ZIP     X         Y
27513 1.37528 0.62492
27013 1.40808 0.62387
```

After running PROC GINSIDE, the CUSTOMERS data set has a REGION_ID variable. Observations with X and Ys that are inside a region have that region's value. In our case, the sales region ID is 999999. If the point was not in any of the map areas, the value is missing (a dot in SAS).

```
ZIP     X         Y      region_id
27513 1.37528 0.62492     999999
27013 1.40808 0.62387        .
```

Now that you know which customers are inside the sales region, you can do something special in their Annotate code to make them stand out in the map. In this case, I am making them red, and I am also using a different text character for their marker ('25cb'x is a circle character of the `albany amt/unicode` font). In the past, you might have used the W character of the SAS MARKERE font, but the Albany AMT characters render with nice smooth edges and look much better.

Note: The Albany AMT font is shipped with SAS 9.2 and higher, and is automatically available on all platforms.

```
data customers; set customers;
if (region_id ne .) then do;
 style='"albany amt/unicode"';
 size=2.9;
 text='25cb'x;
 color="red";
 end;
run;
```

So, here is what all the Annotate pieces (customer markers and the sales region outline) look like at this point:

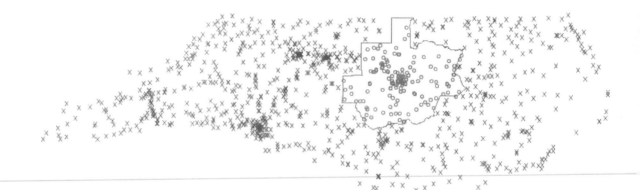

Now that you have all the Annotate pieces ready, you can combine them with the map, project the latitude and longitude coordinates, and then separate the map and Annotate data sets (as explained in more detail in Example 25, "Plotting Multiple Graphs on the Same Page").

```
data mymap; set maps.counties (rename=(state=num_state)
where=(num_state=stfips('NC')));
 state=fipstate(num_state);
run;

data combined; set mymap customers sales_region;
run;

proc gproject data=combined out=combined dupok;
  id state county;
run;

data mymap customers sales_region; set combined;
  if anno_flag=1 then output customers;
  else if anno_flag=2 then output sales_region;
  else output mymap;
run;
```

And finally, you are ready to use PROC GMAP to draw the NC county map, and annotate the customer markers and sales region outline. I specify the two Annotate data sets in two different ANNO= options for convenience (since PROC GMAP enables that). Alternatively, you could combine the two Annotate data sets, and specify the combined data set in just one location.

```
pattern1 v=msolid c=&landcolor;

title1 "Customers Within Multi-County Sales Region";
proc gmap map=mymap data=mymap anno=customers;
 id state county;
 choro state / levels=1 nolegend coutline=grayaa anno=sales_region;
run;
```

Customers Within Multi-County Sales Region

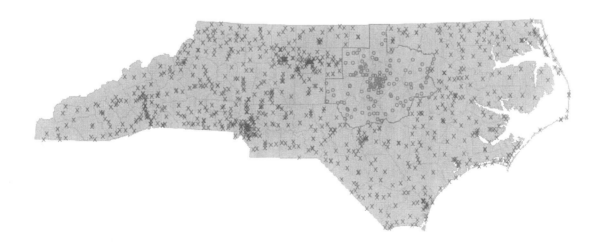

EXAMPLE 9

Determine Customers Within Circular Radius

Purpose: Demonstrate how to use SAS/GRAPH PROC GINSIDE to determine whether customers are in a circular sales area of a given radius.

This example is short and sweet, really just a case of re-using techniques described in detail in two previous examples. This demonstrates that once you learn a "trick" you can apply it in other ways.

First, I get the longitude and latitude (in degrees) of the approximate center of my sales region and store that in a SAS data set. In this example I am grabbing the centroid of the ZIP code 27513 from the sashelp.zipcode data set, but you could get a center longitude and latitude any way that is convenient for you.

```
proc sql;
 create table sales_region as
 select zip, city, x*-1 as long, y as lat
 from sashelp.zipcode
 where (zip eq 27513);
quit; run;
```

Then create a data set containing the X and Y coordinates of a circle that has an n mile radius from that center point. This technique is described in more detail in Example 24, "Plotting Coverage Areas on a Map."

```
%let radius_miles=40;
data sales_region; set sales_region;
  retain xsys ysys '2' anno_flag 2 when 'a';
  length text function $8 style color $20 text $25;
  region_id=999999;
  x=atan(1)/45 * long;
  y=atan(1)/45 * lat;
  d2r=3.1415926/180;
  r=3958.739565;
  xcen=long;
  ycen=lat;
  do degree=0 to 360 by 5;
     if degree=0 then do;
        function='poly';
        style='empty';
        line=1;
        end;
     else do;
```

```
        function='polycont';
        color="&inside_color";
        end;


y=arsin(cos(degree*d2r)*sin(&radius_miles/R)*cos(ycen*d2r)+cos(&radius
_miles/R)*sin(ycen*d2r))/d2r;

x=xcen+arsin(sin(degree*d2r)*sin(&radius_miles/R)/cos(y*d2r))/d2r;
    x=atan(1)/45*x;
    y=atan(1)/45*y;
    output;
  end;

run;
```

The rest of the code is the same as Example 8, "Determine Customers Within Map Borthders," and here is the resulting map:

Customers Within 40 Mile Radius of 27513

EXAMPLE 10

Custom Box Plots

Purpose: Use data-driven/generalized Annotate facility code to draw a custom horizontal box plot.

It is easy to create vertical box plots (also known as box-and-whisker plots) with the SAS/GRAPH GPLOT procedure, using "`symbol interpol=box`". But there is no built-in support for *horizontal* box plots in PROC GPLOT. This is another situation where custom SAS/GRAPH programming comes in handy.

In this example, I demonstrate how to take the pre-summarized values describing a box plot, and create a custom horizontal box plot using the Annotate facility. This code should be easily re-usable for creating similar box plots with your own data, and it should also serve as a good starting point for you to write your own variations of custom box plots.

First, you will need to summarize your data, so that you know the median as well as the upper and lower values for the whiskers and box.

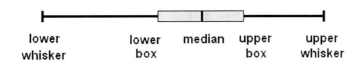

| lower | lower | median | upper | upper |
| whisker | box | | box | whisker |

There are many techniques to calculate summary statistics in SAS; use whichever you are most comfortable with. The end result needs to be a SAS data set containing five numeric values for each "box," arranged as follows:

```
data repdata;
input Name $ 1-20 lower_whisker lower_box Median upper_box
upper_whisker;
datalines;
Allison, R              0.5 6.1  8.2 11.0 17.5
Jeffreys, E             1.0 5.8  9.5 13.0 20.2
Peterson, B             3.0 8.8 11.0 13.1 17.2
Proctor, L              2.0 6.1  9.0 11.5 16.0
Taylor, C               1.5 5.5  9.0 12.6 18.7
Davis, J                2.5 6.0  9.3 11.8 14.8
;
run;
```

To make the code easily re-usable, I define some macro variables for properties of the graph that you might want to change. This enables you to modify the look of the box plot in one central location. It should be obvious what these variables affect based on their mnemonic names (for example, BOXWIDTH is the distance from the vertical center of the box to the outside edge of the box, in the Y-direction).

```
%let boxwidth=1.5;
%let linewidth=.5;
%let linecolor=black;
%let boxcolor=cxCAE1FF;
```

This next DATA step turns the summary data into an annotate data set. Rather than interspersing my description with the code, I will give the description of the entire DATA step first, and then show the code (with graphic examples showing what each piece of the code does).

Note that as in most of my customized graphs, I use the data's coordinate system (xsys/ysys = '2') so that my annotations will line up with the data values. Since the Y-values are character, you need to name your variable YC rather than the usual Y in the annotate data set.

First, I draw the main horizontal line by extending out to the length of the whiskers using the annotate move/draw commands. Likewise, I draw the ends on the whiskers (vertical line at each end of the horizontal line). Note that when I need to move slightly in the Y direction, I change to ysys='7' (relative % coordinate system), and then move the BOXWIDTH down and up. For example, I move down one BOXWIDTH (that is, y=-1*&boxwidth), and then from that lower location I move up two BOXWIDTHs (that is, y=+2*&boxwidth).

Then I draw the colored box in the middle of the line. To draw a box, you first move to the lower-left coordinate, and then you draw a bar to the upper right. You can control the look of the box, using the LINE= and STYLE= variables. In this case, I first draw a solid box of the desired box color, and then draw an outline around the box using an empty box colored with the line color. And last, I draw a vertical line at the median using MOVE and DRAW annotate commands. Note that the use of the WHEN='A' function forces the annotated graphics to be drawn "after" (or on top of) the regular GPLOT plot markers.

```
data repanno; set repdata;
xsys='2'; ysys='2'; hsys='3'; when='a';
length function color $8 style $15;

size=&linewidth;

/* draw the long/horizontal Whisker line */
color="&linecolor";
ysys='2'; yc=Name;
function='move'; x=lower_whisker; output;
function='draw'; x=upper_whisker; output;
```

```
/* draw the vertical ends on the whiskers */
color="&linecolor";
ysys='2'; yc=Name;
function='move'; x=lower_whisker; output;
function='move'; ysys='7'; y=-1*&boxwidth; output;
function='draw'; line=1; y=+2*&boxwidth; output;
ysys='2'; yc=Name;
function='move'; x=upper_whisker; output;
function='move'; ysys='7'; y=-1*&boxwidth; output;
function='draw'; line=1; y=+2*&boxwidth; output;
```

```
/* draw the Box, using annotate 'bar' function */
color="&boxcolor";
ysys='2'; yc=Name;
function='move'; x=lower_box; output;
function='move'; ysys='7'; y=-1*&boxwidth; output;
function='bar'; line=0; style='solid'; x=upper_box;
y=+2*&boxwidth; output;
```

```
/* draw an empty black outline box, around the solid blue box */
color="&linecolor";
ysys='2'; yc=Name;
function='move'; x=lower_box; output;
function='move'; ysys='7'; y=-1*&boxwidth; output;
function='bar'; line=0; style='empty'; x=upper_box; y=+2*&boxwidth;
output;
```

```
/* Median Line */
color="&linecolor";
ysys='2'; yc=Name;
function='move'; x=Median; output;
function='move'; ysys='7'; y=-1*&boxwidth; output;
function='draw'; line=1; y=+2*&boxwidth; output;
run;
```

Once you have created the Annotate data set, all you have to do is use PROC GPLOT to draw some empty axes and annotate your custom graphics onto the plot. In this case, I plot the median values and make the plot markers very small and the same color as the background, to get the empty axes.

```
goptions gunit=pct htitle=5 ftitle="albany amt/bold" htext=3.0
ftext="albany amt" ctext=gray33;

axis1 label=none value=(justify=right) offset=(8,8);

axis2 order=(0 to 25 by 5) minor=none label=none offset=(0,0);

title1 ls=1.5 "Annotated Horizontal Boxplot";
title2 "SAS/GRAPH Gplot";

symbol1 value=none interpol=none height=.001 color=white;
proc gplot data=repdata anno=repanno;
 plot name*Median /
 vaxis=axis1 haxis=axis2
 autohref chref=graydd;
run;
```

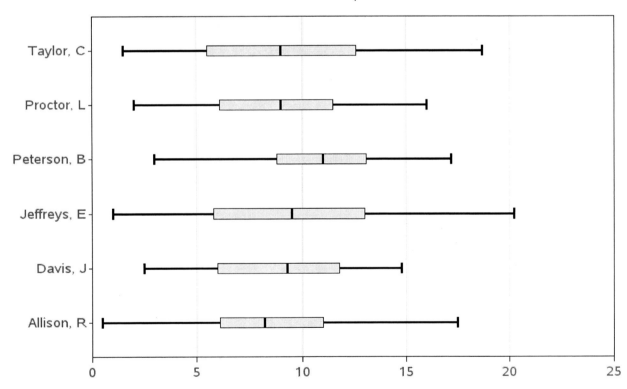

EXAMPLE 11

Using Company Logos in Graphs

Purpose: Put a company logo at the top of a graph (as well as a few other tricks!).

I often get my inspiration for new graphs from real life. For example, when my electric utility company started printing a little bar chart on my monthly power bill, I thought that was a pretty useful thing. I decided to see if I could do the exact same graph with SAS/GRAPH. It wasn't quite as easy as I first thought, and now I am even including it in my book of *tricky* examples.

In this example, we need 13 months of power consumption values. Here is the data we will work with:

```
data my_data;
format kwh comma5.0;
format date mmyy5.;
input date date9. kwh;
datalines;
15sep2002    800
15oct2002    550
15nov2002    200
15dec2002    190
15jan2003    250
15feb2003    200
15mar2003    225
15apr2003    190
15may2003    325
15jun2003    350
15jul2003    675
15aug2003    775
15sep2003    875
;
run;
```

Why 13, rather than 12, you might ask? By showing 13 months, it allows you to not only see the 12-month trend, but also to compare the current month to the same month a year ago. Using some minimal code, we can easily plot this data, and get a basic bar chart:

```
goptions gunit=pct htitle=6 ftitle="albany amt/bold" htext=4.5
ftext="albany amt/bold";

title1 ls=1.5 "kWh Usage History";
pattern1 v=s c=graydd;
axis1 label=none offset=(5,5) value=(angle=90);
axis2 label=none minor=none major=(number=5);

proc gchart data=my_data;
vbar date / discrete type=sum sumvar=kwh
 maxis=axis1 raxis=axis2 noframe
```

```
    autoref cref=graydd clipref;
run;
```

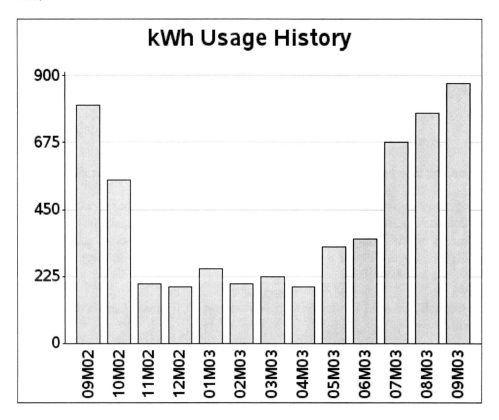

Although this is a reasonably good graph, it still does not look like the graph in my monthly electric bill. This is where the customizing comes into play.

First, let's make the current month have a different fill color and pattern than the other months. This way it is easy to see which one is the current (that is, the most important) month. You can do this by creating a variable (BILLMONTH) and assigning it one value for the current month and a different value for all previous months, and then specify that variable in PROC GCHART's SUBGROUP= option. I am using crosshatch patterns (x1 and x2) to fill the bars, since this is the way they were in my power bill. Some consider crosshatching a bit "old school," but it is probably preferable to shades of gray when using low-quality black-and-white printing.

```
%let targetdate=15sep2003;

data my_data; set my_data;
if date eq "&targetdate"d then billmonth=1;
else billmonth=0;
run;

pattern1 v=x1 c=graybb;
pattern2 v=x2 c=black;
proc gchart data=my_data;
vbar date / discrete type=sum sumvar=kwh
 subgroup=billmonth nolegend
 maxis=axis1 raxis=axis2 noframe
 autoref cref=graydd clipref;
run;
```

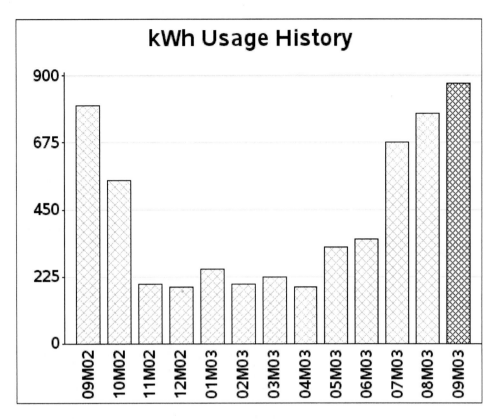

And for the next enhancement, we need to print the customer's address in the top-left area of the page. There are several ways to do this using notes, annotation, and so on, but I think the easiest way is to use title statements. To get the spacing right, I first put a blank title at the top, followed by several left-justified titles (with the three lines of the address), followed by another blank title for spacing, and then a final (centered) title to label the graph. Notice that I am specifying justification (such as `j=l` for left-justified) and plain (non-bold) fonts for the address text. Otherwise, they would default to being centered and bold (since my FTEXT is using a bold font).

```
title1 h=8 " ";
title2 j=l h=4 f="albany amt" " JOHNNY LIGHTNING";
title3 j=l h=4 f="albany amt" " 123 TREADMILL WAY";
title4 j=l h=4 f="albany amt" " CARY NC 27513";
title5 h=6 " ";
title6 j=c h=6 "kWh Usage History";
{same gchart code as before}
```

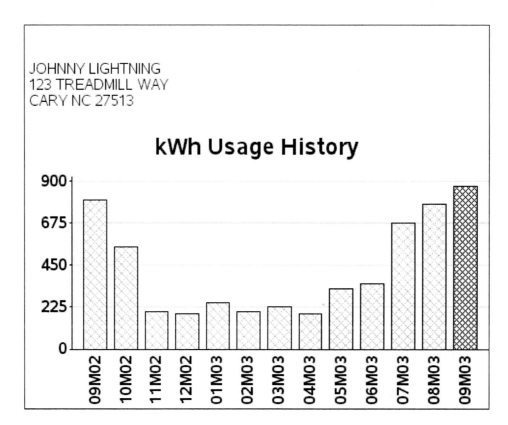

And now for the monthly bar labels: the built-in SAS mmyy5. format works fine, but is not so user-friendly to the person reading the graph. Also, the bill from my electric company used three-character text month names, and my goal was to create a graph like that one. We would normally use the monname3. format that is built into SAS, but that causes some undesirable interactions (because with 13 months, two of them have the same name). Therefore, we have to get a little tricky.

In this case, we will hardcode the text values for the bars, via the value= option of the axis statement. But rather than hardcoding the monthly values directly (which would have to be re-hardcoded each month), we will create these values from the data. The following code loops through the data, appending the month name to the value list each time through, for each of the 13 months (note that the RETAIN statement is important here!). The 't= specifies which tick mark (that is, bar), and the text in quotes is the text value to print under that bar. Since the graph in my electric bill leaves out every other month's text value (so the axis is not cluttered), I make the value blank for every other bar, using some trickery with mod(_n_,2) to determine which bars are odd and even.

```
data my_data; set my_data;
length axis_text $200;
retain axis_text;
if mod(_n_,2)=1 then
 axis_text=trim(left(axis_text))||' t='||trim(left(_n_))||'
'||quote(put(date,monname3.));
else
 axis_text=trim(left(axis_text))||' t='||trim(left(_n_))||' '||quote('
');
run;
```

Here is a portion of the resulting data set. Note that axis_text gets longer with each observation.

```
Obs kwh date  axis_text
 1  800 09M02 t=1 "Sep"
 2  550 10M02 t=1 "Sep" t=2 " "
 3  200 11M02 t=1 "Sep" t=2 " " t=3 "Nov"
 4  190 12M02 t=1 "Sep" t=2 " " t=3 "Nov" t=4 " "
{and so on...}
```

I then convert the final value—the value of the last (MAX) date—into a macro variable, so that I can use it in the axis statement. I use AXIS_TEXT as the name of the macro variable (same as the variable name in the data set).

```
proc sql;
select axis_text into :axis_text from my_data having date=max(date);
quit; run;

axis1 label=none offset=(5,5) value=( &axis_text );
```

Now the bars are labeled with the three-character month name, and every other one is blank. Note that the first and last bars are both labeled as "Sep" and that causes no problem using this technique.

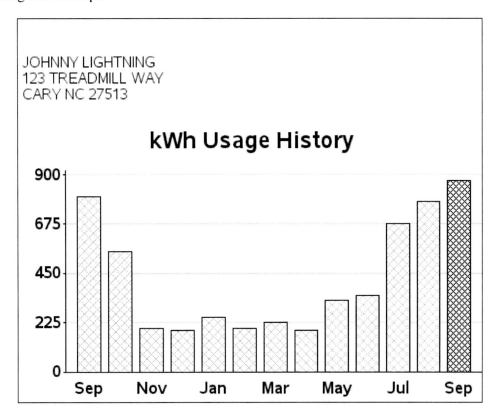

For the finishing touch, we need a power company logo at the top of the page! You can do this fairly easily using two annotate commands. First, use function='move' to move the x and y coordinates of the bottom-left corner where you want the image, and then use function='image' specifying the coordinates of the top-right corner. See Example 12,

"Annotated Map Borders," for another example of annotating an image. Note that this time I am positioning the image in the blank space created by the title1. If you need more blank space for a larger logo, you can increase the height of the blank title1.

```
data logo_anno;
   length function $8;
   xsys='3'; ysys='3'; when='a';
   function='move'; x=0; y=92; output;
   function='image'; x=x+36.4; y=y+7; imgpath='power_logo.png';
style='fit'; output;
run;

proc gchart data=my_data anno=logo_anno;
vbar date / discrete type=sum sumvar=kwh
 subgroup=billmonth nolegend
 maxis=axis1 raxis=axis2 noframe
 width=5 space=2.25 coutline=black
 autoref cref=graydd clipref;
run;
```

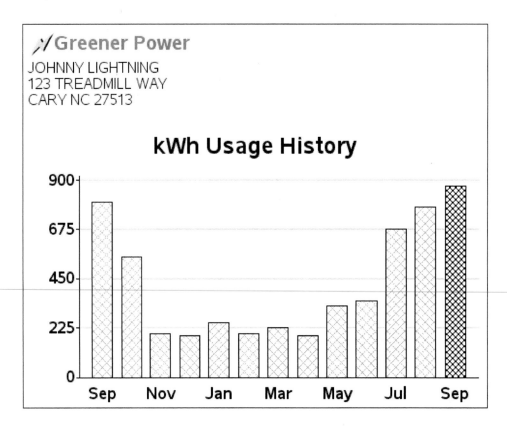

EXAMPLE 12

Annotated Map Borders

Purpose: This example demonstrates how to draw darker state outlines on a U.S. county map. The technique can be easily adapted to any map that has geographical boundaries at a lower level in which you want to also show groupings at a higher level. (This technique was adapted from SAS Technical Support's example number 24904.)

There is no built-in way to represent two different levels of map borders in SAS/GRAPH. Therefore, you have to create a custom map using the Annotate facility. In a nutshell, you create a normal county map, and then annotate the state outline on top of it using a line that is darker or thicker or both.

You might be tempted to try combining maps.uscounty and maps.us, but those two do not necessarily line up perfectly. In order to guarantee that the state outline lines up exactly with the counties, you will need to create the state outline from the exact same points as the county outlines. You do this by removing the internal county boundaries within the states (using PROC GREMOVE), leaving just the state outlines.

The only requirement is that you start with a map that has a "compound ID" (that is, two variables to uniquely identify each area in the map, such as ID STATE COUNTY). First you will want to make a copy of that map, and you can use a simple DATA step for that. When you make a copy of the map, this gives you an easy opportunity to reduce the density of the map, or subset it to exclude unwanted areas, and so on.

```
data uscounty;
   set maps.uscounty;
run;
```

This is what the USCOUNTY data set looks like, when plotted with PROC GMAP. Note that there are no state outlines; therefore, it is difficult to know where one state ends and another begins.

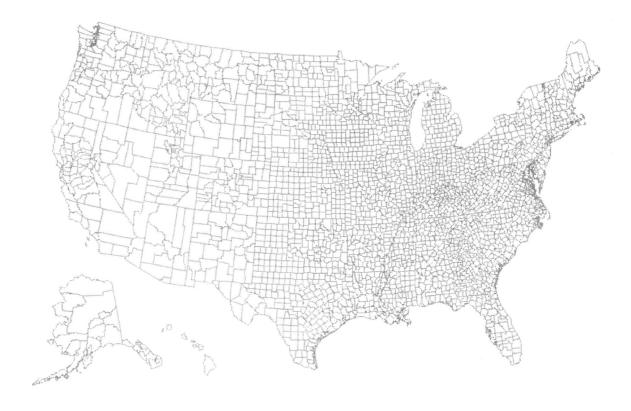

The next step is to create a data set containing just the X and Y points of the outermost boundaries of the states (that is, remove all the internal boundaries). SAS has a procedure specifically for that purpose. Just make sure that you put the outermost border's variables in the BY statement, and the innermost border's variable in the ID statement.

```
proc gremove data=uscounty out=outline;
  by STATE;
  id COUNTY;
run;
```

This is what the OUTLINE data set looks like:

At this point, you now have a county map and a state outline map. You *could* save the GRSEG entries, run PROC GREPLAY on them so that they are on top of each other, and you would probably achieve the desired result (that is, a county map, with darker state outlines). But there are several factors that might affect the positioning of the maps within the GRSEG (such as height, length, and auto-sizing of titles, legends, and so on). They could easily cause the state map borders to not line up exactly with the county map. At the very least this could cause the map to look bad, and at the very worst it could cause the map to be misinterpreted, thus compromising the data integrity.

By comparison, using Annotate to draw the state outlines *guarantees* the state borders will line up correctly with the county borders.

This next DATA step is the only part of this example that is complex. It converts the outline map data set into an Annotate data set. Basically, for each state you move to the first X and Y coordinate, then draw from coordinate to coordinate until you are done with that state, and then move (that is, "lift the pen and stop drawing") to the next state. If the state is broken into multiple segments, then each segment of the state is handled separately. That is why the segment variable is also included in the BY statement.

Fortunately, you do not necessarily have to understand how this code works in order to use it. The only things you might want to edit here are the BY STATE SEGMENT (if your map uses a different variable other than STATE), and the COLOR and SIZE variables (in case you want to use a color other than gray33, or if you want to make the line thicker).

```
data outline;
   length COLOR FUNCTION $ 8;
   retain XSYS YSYS '2' COLOR 'gray33' SIZE 1.75 WHEN 'A' FX FY
FUNCTION;
   set outline; by STATE Segment;
   if first.Segment then do;
      FUNCTION = 'Move'; FX = X; FY = Y; end;
   else if FUNCTION ^= ' ' then do;
      if X = .  then do;
         X = FX; Y = FY; output; FUNCTION = ' '; end;
      else FUNCTION = 'Draw';
   end;
   if FUNCTION ^= ' ' then do;
      output;
      if last.Segment then do;
         X = FX; Y = FY; output; end;
   end;
run;
```

Here is a little extra explanation about the Annotate commands. I use `xsys='2'` and `ysys='2'` so the annotation uses the same coordinate system as the map. This guarantees that the annotated X and Y coordinates line up exactly with the map's X and Y coordinates. `When='A'` causes the annotation to be drawn after the map; therefore, it will be visible on top of the map (as opposed to `when='B'`, which would draw the annotation before, or behind, the map). The values of FX and FY (first-X and first-Y) are *retained*, so you can connect the last point of each segment back to the first point.

All the hard work is basically done now. All you have to do is put it to use. For the purpose of keeping the example code small and easy to follow, I first demonstrate how to create the map using just a few observations of response data.

```
data combined;
input state county bucket;
datalines;
37  37 1
37   1 2
51 117 3
37 125 4
37  51 5
;
run;
```

I usually hardcode a few options to control the colors and fonts in order to override any ODS style that might be in effect. Note that I am using the new TrueType fonts that ship with SAS 9.2 (these will work on Windows, UNIX, or mainframe). I specify five pattern statements because I have divided my data into five "buckets."

```
goptions cback=white;
goptions gunit=pct htitle=4 ftitle="albany amt/bold" htext=3
ftext="albany amt";

pattern1 v=s c=cxD7191C;
pattern2 v=s c=cxFDAE61;
pattern3 v=s c=cxFFFFBF;
pattern4 v=s c=cxA6D96A;
pattern5 v=s c=cx1A9641;
```

And finally, I use PROC GMAP to draw the county map and annotate the state outlines. Note that the `coutline=graydd` controls the outline color of the counties, and the outline color of the states is controlled by the COLOR= variable in the Annotate data set.

```
title1 "U.S. County map, with Annotated State Borders";
proc gmap data=combined map=uscounty all anno=outline;
id state county;
choro bucket /
 midpoints = 1 2 3 4 5
 coutline=graydd
 nolegend;
run;
```

U.S. County map, with Annotated State Borders

One thing worth mentioning is that I use the MIDPOINTS= option to guarantee that the legend will have all five possible bucket values, even if the particular data I am plotting does not have all five values in it. This guarantees that the desired data values will always have the desired legend color, and you can easily compare maps.

In order to generate the final map as shown in the example on the author's Web page, you will need the full set of response data (which you can download from the Web page), and you will also need a little more code for the titles and the way the legend is displayed.

In the titles, I make the words "out of" red and "into" green in order to reinforce the color scheme used in the map legend. This is accomplished by breaking the title text into several pieces, and specifying the color before each piece of the title, using C=. I also use the

LINK= option before the "IRS" text, so I can add a link to the IRS Web site where I got the data, and I make that text blue so people will know it has a link. Note that links in the title, as with most other links, will generally work only with ODS HTML output.

```
title1 "Net Effect of People moving "  c=cxD7191C "out of" c=gray55 "
& " c=cx1A9641 "into";
title2 " &countynm County, &statecode";
title3 "between years 20&year1 and 20&year2 "
       "(data source: " c=blue
link="http://www.irs.gov/taxstats/article/0,,id=212695,00.html" "IRS"
c=gray ")";
```

Where did I get the text values for the macro variables used in TITLE2? I queried them out of some data sets supplied by SAS. At the top of the SAS job, I specify the numeric FIPS code for the state and county of interest, and then I use PROC SQL to look up the text county name in the maps.cntyname data set, and stuff that value into the COUNTYNM macro variable. Similarly, I look up the state abbreviation in maps.us. Note that I use separated by ' ' to guarantee that the values of the macro variables do not get padded with blanks (which would throw off the spacing in the titles).

```
%let state=37;
%let county=183;

proc sql;
select propcase(countynm) into :countynm separated by ' '
 from maps.cntyname where state=&state and county=&county;
select unique statecode into :statecode separated by ' '
 from maps.us where state=&state;
quit; run;
```

If you were to plot the map with what we have got so far, it would look something like this:

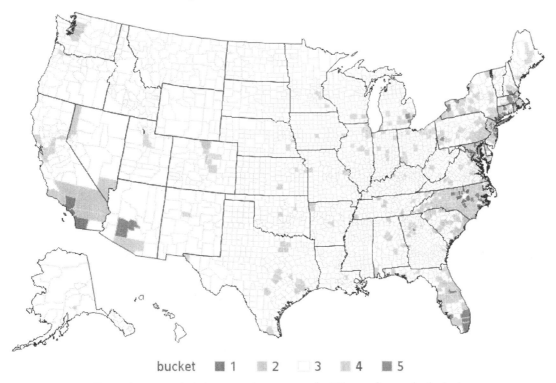

Net Effect of People moving out of & into
Wake County, NC
between years 2007 and 2008 (data source: IRS)

bucket ■ 1 ■ 2 □ 3 ■ 4 ■ 5

For privacy, only inter-county moves of >10 people are included.

Although I have assigned each county a "bucket" value of 1 through -5, I actually want them to show up as something a little more descriptive in the legend. Therefore, I create a user-defined format so the numeric values 1 through -5 will show up as descriptive text. Note that I use macro variables to hold the numeric cut-off values for the buckets. That way, I can use the macro variable in both the IF statement (when assigning the bucket values) and the user-defined format, guaranteeing that they will match. Otherwise, someone editing the program might change the value in one place, and not the other, which would compromise the data integrity.

```
proc format;
value buckets
1 = "< &bucketn"
2 = "&bucketn to 0"
3 = "0"
4 = "0 to &bucketp"
5 = "> &bucketp"
;
run;
```

Note that there are several ways to control the binning and the appearance of the legend. Just consider this as one example (and not necessarily the best technique for every situation).

Although the title indicates which county is the subject of the map, I thought it would also be good to indicate that graphically, so I annotate a "star" in the center of that county. I just approximate the geographical center of the county by taking the average X and Y from maps.uscounty. I use the SAS/GRAPH MARKER font "V" character to draw a star, and then the MARKERE (e=empty) font to draw an outline around the star and make it more visible.

```
proc sql;
create table anno_star as
select avg(x) as x, avg(y) as y
from maps.uscounty
where state=&state and county=&county;
quit; run;

data anno_star; set anno_star;
length function style color $8;
xsys='2'; ysys='2'; hsys='3'; when='a';
function='label';
text='V';
size=3.0;
style='marker'; color='cyan'; output;
style='markere'; color='black'; output;
run;
```

The bucket's user-defined format is applied via the FORMAT statement, and the annotated star (ANNO_STAR) is specified in the second ANNO= option.

```
proc gmap data=combined map=uscounty all anno=outline;
id state county;
format bucket buckets.;
choro bucket / discrete
 midpoints = 1 2 3 4 5
 coutline=graydd
 legend=legend1
 anno=anno_star;
run;
```

And here is what the final map looks like:

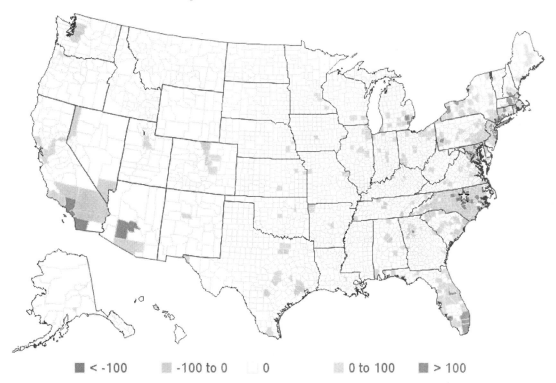

Net Effect of People moving out of & into
Wake County, NC
between years 2007 and 2008 (data source: IRS)

■ < -100   ■ -100 to 0   □ 0   ■ 0 to 100   ■ > 100

For privacy, only inter-county moves of >10 people are included.

EXAMPLE 13

Population Tree Charts

Purpose: Demonstrate how to create a population tree chart with the axis in the middle.

Population tree charts (also known as, paired bar charts or sometimes tornado charts) are very useful when comparing two populations by age group.

Most savvy SAS/GRAPH programmers have found ways to create a basic population tree chart with a minimum of custom programming. For example, you might make the values for one population negative, to have them plot on the left side of the response axis. For a nice touch, you could create a user-defined format to make the negative response axis values look positive. A paper I co-authored with Dr. Moon W. Suh for the 1994 SouthEast SAS Users' Group entitled "Creating Population Tree Charts (Using SAS Graph Software") describes this technique in detail.[1] Here is an example:

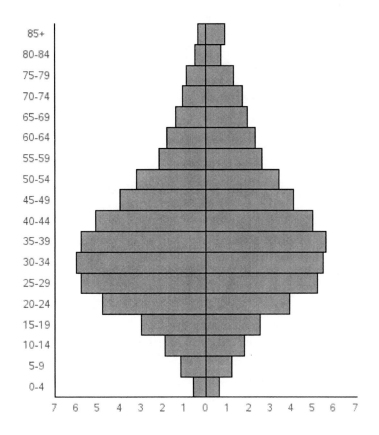

The graph is *not bad*, but I think it would look a lot better to have the midpoint axis values *between* the left and right halves rather than on the side. That way it is easier to see which label goes with which bar, and it makes the graph move visually symmetrical.

A good example of the type of layout can be found in the following Census publication: http://www.census.gov/prod/2003pubs/c2kbr-34.pdf.

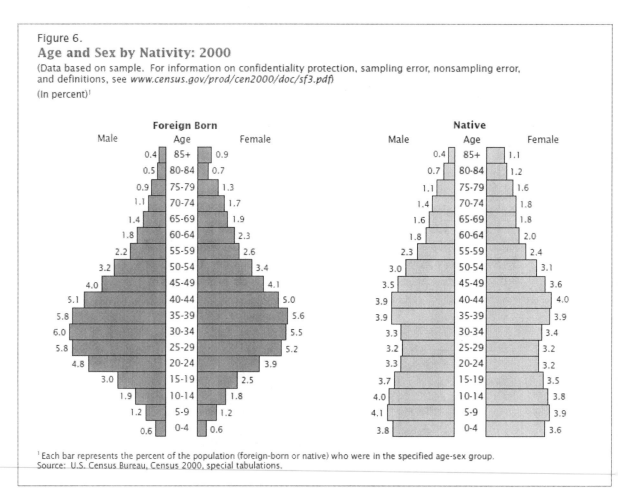

Figure 6.
Age and Sex by Nativity: 2000
(Data based on sample. For information on confidentiality protection, sampling error, nonsampling error, and definitions, see *www.census.gov/prod/cen2000/doc/sf3.pdf*)
(In percent)[1]

[1] Each bar represents the percent of the population (foreign-born or native) who were in the specified age-sex group.
Source: U.S. Census Bureau, Census 2000, special tabulations.

But this type of layout is a bit tougher to do in SAS/GRAPH, and requires some custom SAS programming.

Basically, the approach involves creating each half of the chart separately, and then using PROC GREPLAY to place the two halves side by side.

Let's start with the chart for the foreign-born population. First we will need some data. For each pair of bars (left and right pieces), we will need a label for the age group and the male and female populations. While I am creating the data set, I make the male value negative (because that bar will be going to the left), and I assign a DATA_ORDER variable so I can later plot the bars in that same order. This will be important later, because we do not want to plot the bars alphabetically or ascending or descending.

```
data foreign;
input age_range $ 1-5 male female;
male=-1*male;
data_order=_n_;
datalines;
```

```
85+   0.4 0.9
80-84 0.5 0.7
75-79 0.9 1.3
70-74 1.1 1.7
65-69 1.4 1.9
60-64 1.8 2.3
55-59 2.2 2.6
50-54 3.2 3.4
45-49 4.0 4.1
40-44 5.1 5.0
35-39 5.8 5.6
30-34 6.0 5.5
25-29 5.8 5.2
20-24 4.8 3.9
15-19 3.0 2.5
10-14 1.9 1.8
 5-9  1.2 1.2
 0-4  0.6 0.6
;
run;
```

The DATA_ORDER variable is going to control the ordering of the bars (that is, we will use the discrete DATA_ORDER values as our bar midpoints), but we do not want those numeric values to show up as the bar labels. Instead, we want the AGE_RANGE text to show up. Therefore, we create a user-defined format, so the numeric DATA_ORDER values show up as the AGE_RANGE text.

```
data control; set foreign (rename = ( data_order=start
age_range=label));
 fmtname = 'my_fmt';
 type = 'N';
 end = START;
run;


proc format lib=work cntlin=control;
run;
```

Let's start with the right side of the chart (the female population). You can easily produce the basic bars using the following code. There is nothing tricky about that besides the user-defined format for the midpoint labels:

```
pattern v=s c=cx6782b1;

axis3 label=none value=(justify=center);
axis4 label=none minor=none major=none value=(color=white) style=0
order=(0 to 7 by 1);

goptions xpixels=186 ypixels=435;

proc gchart data=foreign;
 format data_order my_fmt.;
 hbar data_order / discrete
 type=sum sumvar=female
 nostats noframe
 coutline=black space=0
 maxis=axis3 raxis=axis4
 name="f_female";
run;
```

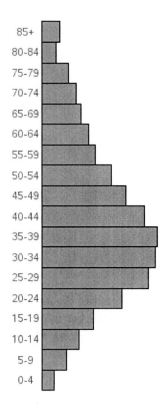

Now, to get the text labels and such around the bar chart, I will use some annotation. First, let's create the labels above the graph ("Foreign Born", "Age", and "Female"). These are just standard annotated labels with the positions hardcoded using xsys and ysys='3' (that is, the X and Y positions are specified as a percentage of the graphics output area).

```
data foreign_right1;
  length text $20;
  function='label';
  xsys='3'; ysys='3'; position='5'; when='a';
  x=8;  y=95.5; text='Foreign Born'; style='"albany amt/bold"'; output;
  x=8;  y=92; text='Age'; style=''; output;
  x=65; y=92; text='Female'; output;
run;
```

Next, we will annotate the data values at the end of each bar. Note that I am using the data values and therefore using the coordinate system (xsys and ysys='2'). Since it is a horizontal bar chart I use the midpoint variable instead of the Y variable. I use position='6' so the text appears to the right of the annotate location, and I even add a little bit of offset in the X direction (x=female+.2). Every variable in this Annotate data set is important in order to get the desired text in the desired position.

```
data foreign_right2; set foreign;
  function='label';
  xsys='2'; ysys='2'; position='6'; when='a';
  x=female+.2; midpoint=data_order;
  text=trim(left(put(female,comma5.1)));
run;
```

I then run the same PROC GCHART code as before, but this time I specify the two Annotate data sets to get the extra text. I could have combined them into one data set, but since PROC GCHART enables me to specify two Annotate data sets, I find it convenient to leave them separate. Note that I have also added a blank title to create some white space at the top of the chart, and make room for the annotated labels.

```
title1 h=7pct " ";

proc gchart data=foreign anno=foreign_right1;
format data_order my_fmt.;
hbar data_order / discrete type=sum sumvar=female
 nostats noframe coutline=black space=0
 maxis=axis3 raxis=axis4 anno=foreign_right2
 name="f_female";
run;
```

Do not worry that part of the "Foreign Born" label is chopped off. There will be more room for it in the final chart after we use PROC GREPLAY.

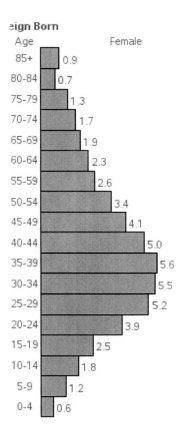

The left half (the male population) is basically the same as the right half except that the data values are negative (so they will point to the left), and the annotated text uses position='4' so it goes to the left of the annotation location. I suppress the midpoint values on the axis (value=none) because I do not need two copies of them. And I make the XPIXELs a little smaller than I used on the right half, because I do not need the extra real estate for the midpoint value labels.

```
data foreign_left1;
 length text $20;
 function='label';
 xsys='3'; ysys='3'; position='5'; when='a';
 x=50; y=92; text='Male'; output;
run;

data foreign_left2; set foreign;
 function='label';
 xsys='2'; ysys='2'; position='4'; when='a';
 x=male-.1; midpoint=data_order;
 text=trim(left(put(abs(male),comma5.1)));
run;

axis1 label=none value=none style=0;
axis2 label=none minor=none major=none value=(color=white) style=0
order=(-7 to 0 by 1);

goptions xpixels=165 ypixels=435;

title1 h=7pct " ";

proc gchart data=foreign anno=foreign_left1;
hbar data_order / discrete type=sum sumvar=male
 nostats noframe coutline=black space=0
 maxis=axis1 raxis=axis2 anno=foreign_left2
 name="f_male";
run;
```

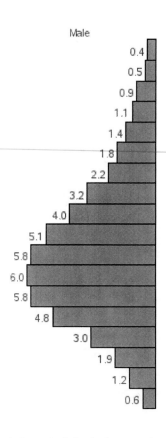

Now that we have created the left and right halves and saved them in GRSEG entries (named using the NAME= option), we can put the two halves together using PROC

GREPLAY.

We will need a custom GREPLAY template for this, which can be created by specifying the X and Y coordinates of the four corners of each area that will contain a piece of the graph. I create panels numbered 1–4 for the left and right halves of the two graphs (four pieces total), and then a fifth panel as the whole area (on which I will use PROC GSLIDE to create the overall title and footnote text for). I name this custom template CENSUS.

```
goptions xpixels=700 ypixels=500;

proc greplay tc=tempcat nofs igout=work.gseg;
  tdef census des='Panels'

    1/llx = 0    lly =  3
      ulx = 0    uly = 90
      urx =24    ury = 90
      lrx =24    lry =  3
    2/llx =24    lly =  3
      ulx =24    uly = 90
      urx =50    ury = 90
      lrx =50    lry =  3

    3/llx =50    lly =  3
      ulx =50    uly = 90
      urx =74    ury = 90
      lrx =74    lry =  3
    4/llx =74    lly =  3
      ulx =74    uly = 90
      urx =100   ury = 90
      lrx =100   lry =  3

    5/llx = 0    lly =  0
      ulx = 0    uly = 100
      urx =100   ury = 100
      lrx =100   lry =  0

  ;
```

Here is what the custom template layout looks like (with the sections numbered).

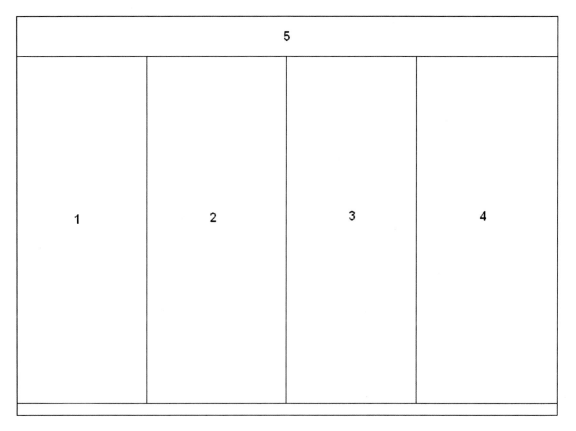

With the following PROC GREPLAY code, you can display the GRSEGs for the left and right halves of the Foreign Born chart (F_MALE and F_FEMALE) into the template. The image below shows the results with the template borders turned on so you can see exactly where the pieces are going:

```
template = census;
treplay
 1:f_male 2:f_female
 ;
run;
```

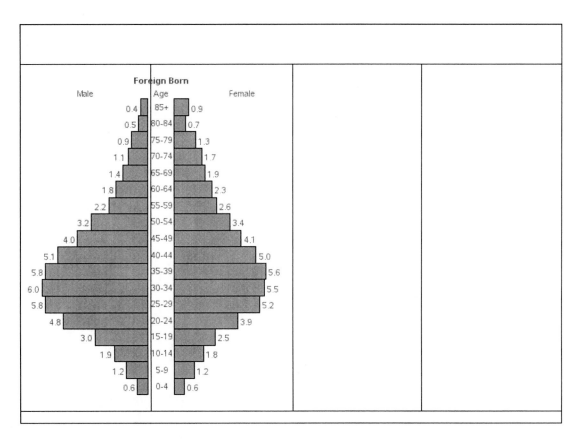

For the overall title and footnote text, we will use a simple PROC GSLIDE to create that.

```
goptions xpixels=700 ypixels=500;

title1 j=l c=cx005096 ls=.8 " Age and Sex by Nativity: 2000";
title2 j=l h=2.4 ls=.4 " (Data based on sample.  For information on
confidentiality protection, sampling error, nonsampling error.";
title3 j=l h=2.4 ls=.4 " and definitions, see " f="albany amt/italic"
"www.census.gov/prod/cen2000/doc/sf3.pdf" f="albany amt" " )";

footnote1 j=l h=2.2 " Each bar represents the percent of the
population (foreign-born or native) who were in the specified age-sex
group.";
footnote2 j=l h=2.2 " Source: U.S. Census Bureau, Census 2000, special
tabulations.";
footnote3 h=1pct " ";

proc gslide name="titles";
run;
```

Creating the left and right sides of the "Native Born" chart are basically the same as for the "Foreign Born" chart, so I will not waste the space by including the code here. (It's in the full program you can download from the author's Web page.) But basically, you create both halves of the graph and name the GRSEGs N_MALE and N_FEMALE. And then you combine the four pieces of the graphs, along with the titles, using the following TREPLAY code in your PROC GREPLAY.

```
treplay
 1:f_male 2:f_female  3:n_male 4:n_female
 5:titles
 ;
run;
```

And that produces the following, which is almost identical to the Census chart.

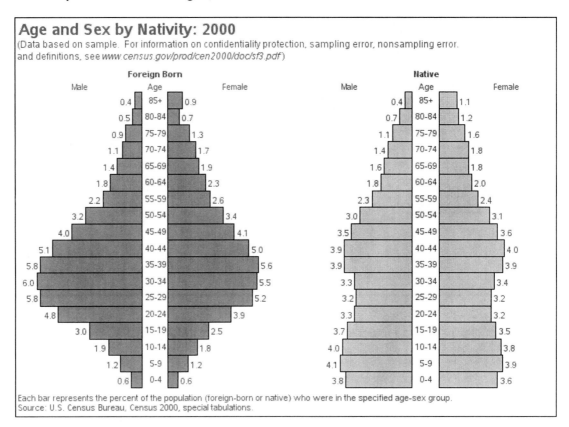

This is a very useful and flexible type of chart. With a little tweaking and customizing, you should be able to re-use this code to plot many different types of data.

Notes

[1] Available at http://support.sas.com/publishing/authors/allison_robert.html.

EXAMPLE 14

Sparkline Table

Purpose: Create a table of custom sparklines.

When you are analyzing and comparing lots of data, it is often useful to see several plots on the same page. Sometimes you might use small multiples, rows and columns of graphs, paneled displays, or even dashboard layouts. For example, here is a multi-graph display from Example 3, "Paneling Multiple Graphs on a Page."

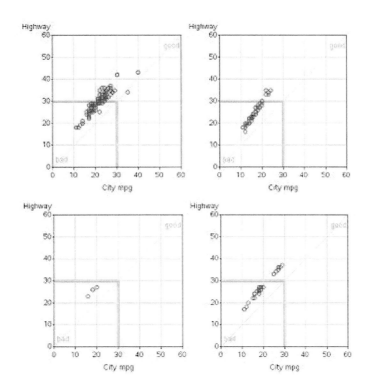

But when you want to pack the maximum number of line charts into the smallest possible space, you will probably want to use *sparklines*. Here is a small sparkline example:

There is not a built-in procedure option to create sparklines using PROC GPLOT, but with a little custom programming, PROC GPLOT can produce sparklines on par with the best of them.

One of the typical uses of sparklines is to show stock market data, and that is what this example demonstrates. The example can be used in two ways: Those who are up to the challenge can use it to learn several very useful tricks to create custom graphs (using both data manipulation and annotation), and the example can also be used as is, to easily create sparkline plots of your own data.

First we will need some data. The data must contain a variable to identify which stock the line represents ("stock" is the text description), and dates and values for the points along the line (in this case we are using the closing price of the stock "close"). The sashelp.stocks data set (which ships with SAS) has data in the needed format, which makes it convenient to use for this example.

I make one slight enhancement. Rather than using only the full-text stock name, I think it is useful to also have the stock ticker abbreviation, so I add a variable named STICKER:

```
data rawdata; set sashelp.stocks (keep = stock date close);
  if stock='Microsoft' then sticker='MSFT';
  if stock='Intel' then sticker='INTC';
  if stock='IBM' then sticker='IBM';
run;
```

So the data I am working with looks like this:

```
stock          sticker        date         close
-------------------------------------------------
IBM             IBM         01DEC2005      $82.20
Intel           INTC        01DEC2005      $24.96
Microsoft       MSFT        01DEC2005      $26.15
{and so on, for other dates...}
```

You can easily plot this data using a simple GPLOT procedure with the three lines overlaid, and this is a good starting point:

```
symbol value=none interpol=join;

axis1 label=none order=(0 to 250 by 50) minor=none offset=(0,0);
axis2 label=none minor=none;

proc gplot data=rawdata;
  format date date7.;
  format close dollar5.0;
  plot close*date=sticker / vaxis=axis1 haxis=axis2;
run;
```

Standard Plot of Data Values

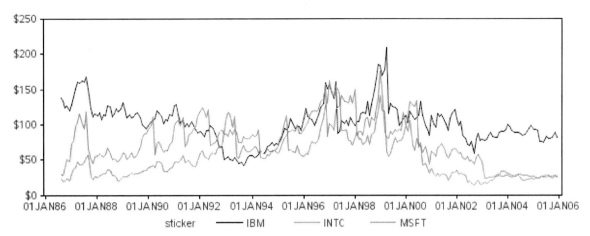

The first steps leading up to the sparkline involve *data manipulation*. First, you will want to "normalize" the values for each line, so that the values are scaled from 0 to 1. One way to do this is to determine the minimum and maximum (MIN_CLOSE and MAX_CLOSE) values of each line using PROC SQL, and then use a DATA step to calculate the normalized closing value (NORM_CLOSE):

```
proc sql noprint;
  create table plotdata as
  select plotdata.*,
   min(close) as min_close, max(close) as max_close
  from plotdata
  group by sticker
  order by sticker, date;
quit; run;

data plotdata; set plotdata;
 norm_close=(close-min_close)/(max_close-min_close);
run;
```

Here is a plot of these normalized values, so you can see the effect this manipulation has made:

```
axis1 label=none order=(0 to 1 by .2) minor=none offset=(0,0);
axis2 label=none minor=none major=none value=none;

proc gplot data=plotdata;
plot norm_close*date=sticker / nolegend
 vaxis=axis1 haxis=axis2;
run;
```

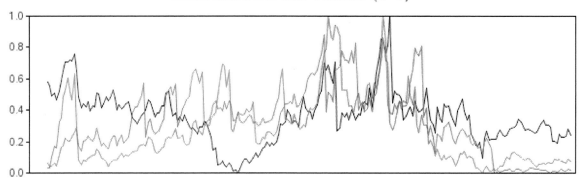

Next adjust the normalized values so they have a 10% spacing above the maximum and below the minimum values. I do this by reducing the values by 80% (which creates 20% spacing above the maximum values), and then adding 10% to shift the values upwards. The end result is 10% spacing above and below. This is something you might want to adjust based on your personal preference and the data you are plotting.

```
data plotdata; set plotdata;
 norm_close=.80*(close-min_close)/(max_close-min_close)+.10;
run;
```

Here is a plot so you can see what effect the changes had on the data. I have added horizontal reference lines so you can easily see the 10% and 90% extents of the modified data.

```
proc gplot data=plotdata;
plot norm_close*date=sticker / nolegend
 vaxis=axis1 haxis=axis2
 vref=.1 .9 cvref=graydd;
run;
```

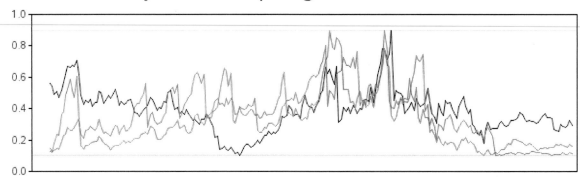

Now for the *key* step that essentially converts the GPLOT into sparklines and offsets each successive line by +1. With the values of each line normalized (0–1) and the values compressed so there is a little spacing above and below, simply adding an offset of +1 (in the Y-direction) to each successive line essentially spreads them out into a sparkline-like layout.

First I number each of the lines (adding a variable called ORDER), and then I use that number to add the offset to each line. I also insert some SAS missing values (norm_close=.) that I will use in conjunction with the SKIPMISS option so each line is not joined with the next successive line. As the name ORDER implies, I use the ORDER variable to control the order of the sparklines. (In this case I am using data order, but I could have sorted the data in some other way before assigning this ORDER variable to arrange the sparklines differently.)

```
data plotdata; set plotdata;
 by sticker notsorted;
 if first.sticker then order+1;
run;

proc sort data=plotdata out=plotdata;
 by sticker date;
run;

data plotdata; set plotdata;
 by sticker;
 norm_close=norm_close+(order-1);
 output;
 if last.sticker then do;
  norm_close=.;
  output;
  end;
run;
```

And you can plot the data to see the progress of the three sparklines thus far using:

```
axis1 label=none order=(0 to 3 by 1) minor=none offset=(0,0);
axis2 style=0 label=none major=none minor=none value=none;

symbol1 value=none interpol=join color=cx00cc00;

proc gplot data=plotdata;
plot norm_close*date=1 / skipmiss
 noframe autovref cvref=graydd
 vaxis=axis1 haxis=axis2;
run;
```

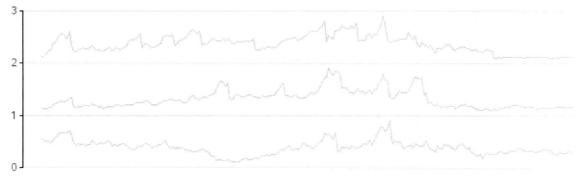

I wanted to show you the above (simplified) code to help you understand how the sparkline offset works, but let's rewind and do that again, this time with a few enhancements.

Many times it is useful to identify the *high* and *low* values on the sparklines (especially with stock market data). Below is the code rewritten to also calculate lines (actually single-point lines) for the minimum and maximum values, and then merge these HIGH_CLOSE and LOW_CLOSE lines back with the main PLOTDATA.

```
proc sql noprint;

  create table high_close as
  select unique sticker, stock, date, norm_close as high_close
  from plotdata group by sticker
  having norm_close=max(norm_close);

  create table low_close as
  select unique sticker, stock, date, norm_close as low_close
  from plotdata group by sticker
  having norm_close=min(norm_close);

  create table plotdata as select plotdata.*, high_close.high_close
  from plotdata left join high_close
  on plotdata.date=high_close.date and
plotdata.sticker=high_close.sticker;

  create table plotdata as select plotdata.*, low_close.low_close
  from plotdata left join low_close
  on plotdata.date=low_close.date and
plotdata.sticker=low_close.sticker;

quit; run;

proc sort data=plotdata out=plotdata;
 by sticker date;
run;

data plotdata; set plotdata;
 by sticker;
 norm_close=norm_close+(order-1);
 high_close=high_close+(order-1);
 low_close=low_close+(order-1);
 output;
 if last.sticker then do;
  close=.;
  norm_close=.;
  low_close=.;
  high_close=.;
  output;
  end;
run;
```

I also calculate a few values from the data to use in the axes. Specifically, I use PROC SQL to calculate the values and save them into macro variables. Then I use those macro variables in the axis statements.

```
proc sql noprint;
 select count(unique(sticker)) into :y_max separated by ' ' from
plotdata;
 select lowcase(put(min(date),date9.)) into :mindate separated by ' '
from plotdata;
 select lowcase(put(max(date),date9.)) into :maxdate separated by ' '
from plotdata;
 select max(date)-min(date) into :days_by separated by ' ' from
plotdata;
quit; run;
```

```
axis1 style=0 label=none major=none minor=none
 order=(0 to &y_max) value=none offset=(0,0);

axis2 style=0 label=none major=none minor=none
 order=("&mindate"d to "&maxdate"d by &days_by) value=none;
```

Below is the code to plot the newly modified data, overlaying the green lines with a blue line (for high values) and red line (for low values). The high and low lines have only one point each, and these are displayed as dots rather than as lines—each controlled by the symbol statements.

```
symbol1 value=none interpol=join color=cx00cc00;      /* green line */
symbol2 value=dot h=5pt interpol=none color=cx42c3ff; /* blue dots -
high value */
symbol3 value=dot h=5pt interpol=none color=cxf755b5; /* red dots -
low value */

proc gplot data=plotdata;
plot norm_close*date=1 high_close*date=2 low_close*date=3 / overlay
skipmiss
 autovref cvref=graydd noframe
 vaxis=axis1 haxis=axis2;
run;
```

Add High/Low Markers

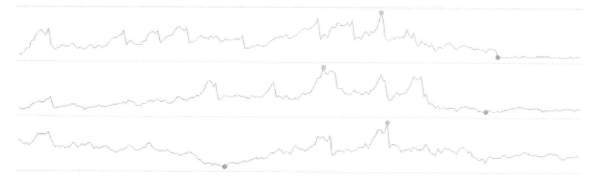

That was a lot of extra work just to display the minimum and maximum values. But the ability to add those subtle touches is what makes custom graphs more useful, and it is what differentiates SAS software from its competitors.

You probably notice that the above lines are a little large for sparklines, given that they stretch all the way across the table. Our next step is to squeeze them in toward the center by adding offset space to the left and right in the axis statement.

```
axis2 style=0 label=none major=none minor=none
 order=("&mindate"d to "&maxdate"d by &days_by)
 value=none offset=(45,40);

{using same Gplot code as before}
```

Add 'Offset' on Left/Right

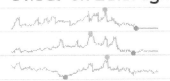

We are now finished with the first phase of the customizations (data manipulations and axis options), and are ready to move on to the "annotation phase."

First we will consider the data-driven annotation. For each line (that is, each stock), we want to annotate some text to the left and right of the line plot in order to show the stock name and ticker, the beginning and ending values, the percent change going from the beginning to the ending value, and finally the minimum and maximum stock values.

I use PROC SQL to create ANNO_TABLE containing all the nonmissing data observations, and then I query the BEGIN_CLOSE and END_CLOSE values for each stock. I then merge those values back in with ANNO_TABLE. Finally I use the SQL procedure's UNIQUE keyword to get just one observation per stock. As with most things in SAS, there are several ways you could accomplish this task. PROC SQL is my personal preference.

```
proc sql noprint;

 create table anno_table as select * from plotdata having
low_close^=.;

 create table begin_close as
  select unique sticker, close as begin_close
  from rawdata group by sticker having date=min(date);
 create table anno_table as select anno_table.*,
begin_close.begin_close
  from anno_table left join begin_close
  on anno_table.sticker=begin_close.sticker;

 create table end_close as
  select unique sticker, close as end_close
  from rawdata group by sticker having date=max(date);
 create table anno_table as select anno_table.*, end_close.end_close
  from anno_table left join end_close
  on anno_table.sticker=end_close.sticker;

 create table anno_table as
  select unique order, sticker, stock, min_close, max_close,
begin_close, end_close
  from anno_table;

quit; run;
```

Now add annotate commands to the ANNO_TABLE data set in order to write out those text strings as annotated labels. I hardcode the X-values based on where I want the "columns" of text, and the Y-values are data-driven. Note that I am using position='3' for left-justified text, and position='1' for right-justified text.

```
data anno_table; set anno_table;
 length text $65;
 function='label'; position='3'; when='a';
 ysys='2'; y=order-1;
```

```
xsys='1';
x=1; text=trim(left(stock)); output;
x=24; text=trim(left(sticker)); output;
position='1';
x=38; text=trim(left(put(begin_close,comma10.2))); output;

x=72; text=trim(left(put(end_close,comma10.2))); output;
x=81; text=trim(left(put((end_close-
begin_close)/begin_close,percentn7.1))); output;
x=90; text=trim(left(put(min_close,comma10.2))); color="cxf755b5";
output;
x=99; text=trim(left(put(max_close,comma10.2))); color="cx42c3ff";
output;
color="";
run;
```

Plot the data using the same code as before, but add `anno=anno_table` to display the annotated text values on the graph:

```
proc gplot data=plotdata anno=anno_table;
plot norm_close*date=1 high_close*date=2 low_close*date=3 / overlay
skipmiss
 autovref cvref=graydd
 vaxis=axis1 haxis=axis2;
run;
```

Annotate Data-Driven Text

Microsoft	MSFT	28.50		26.15	-8.2%	23.70	175.00
Intel	INTC	23.00		24.96	8.5%	13.89	162.25
IBM	IBM	138.75		82.20	-40.8%	42.00	209.19

We have just about got a decent sparkline table now.

We will add one finishing touch. We need some column headings, and some vertical lines between the columns. Let's create a second annotate data set for this. The MOVE and DRAW commands create the lines, the LABEL commands place the text at the top of the columns, and the BAR command fills in the gray area behind the stock-ticker column.

```
data anno_lines;
 length function color $8;
 xsys='1'; ysys='1'; color="graydd";
 when='a';
 function='move';  x=0; y=0; output;
  function='draw'; y=100; output;
 function='move';  x=30; y=0; output;
  function='draw'; y=100; output;

 function='move';  x=82; y=0; output;
  function='draw'; y=100; output;
 function='move';  x=91; y=0; output;
  function='draw'; y=100; output;
 function='move';  x=100; y=0; output;
  function='draw'; y=100; output;

 function='label'; color=''; position='2'; y=100;
 x=35; text="&mindate"; output;
```

```
x=69; text="&maxdate"; output;
x=87; text="low"; output;
x=96; text="high"; output;

when='b';
function='move';  x=23; y=0; output;
  function='bar'; color='grayee'; style='solid'; x=30; y=100; output;
run;
```

I use a blank TITLE2 to create some white space and make room for the column headings, and then specify the second Annotate data set using `anno=anno_lines`.

```
title "Add Column-Headers and Vertical Lines";
title2 height=20pt " ";
proc gplot data=plotdata anno=anno_table;
plot norm_close*date=1 high_close*date=2 low_close*date=3 / overlay
skipmiss
 autovref cvref=graydd
 vaxis=axis1 haxis=axis2
 anno=anno_lines;
run;
```

Add Column-Headers and Vertical Lines

		01aug1986		01dec2005		low	high
Microsoft	MSFT	28.50		26.15	-8.2%	23.70	175.00
Intel	INTC	23.00		24.96	8.5%	13.89	162.25
IBM	IBM	138.75		82.20	-40.8%	42.00	209.19

We have got a really sharp-looking sparkline table now. However, it still has one flaw: it does not show how the data in the lines are scaled. Sparklines do not provide enough space along the Yaxis to show values and tick marks, which let the user know how each line is scaled. Each line could be scaled so that they are all plotted with the minimum Y-axis values being zero. Or they could all be normalized together (that is, normalized against a common minimum and maximum value). Or, as in this case, each line is independently scaled from the minimum to the maximum value.

Therefore, to let the user know how the lines are scaled, I think it is useful to add a footnote:

```
footnote1 j=r c=gray "Note: Each sparkline is independently auto-
scaled.";
```

Stock Prices (sparkline table)

		01aug1986		01dec2005		low	high
Microsoft	MSFT	28.50		26.15	-8.2%	23.70	175.00
Intel	INTC	23.00		24.96	8.5%	13.89	162.25
IBM	IBM	138.75		82.20	-40.8%	42.00	209.19

Note: Each sparkline is independently auto-scaled.

Now for a neat trick: most SAS/GRAPH programmers use the default page size, and let the graph automatically spread out to fill the available space. But with something like a sparkline table, you want the sparklines to stay approximately the same (small) size no matter how many (or few) lines are in the table. Here is a trick you can use to dynamically set the page space based on the number of sparklines. Basically, I calculate a number equal to about 35 pixels per sparkline, plus about 33 pixels for the title and footnote, and use that as the YPIXELS= value.

```
proc sql noprint;
  select 33+35*count(*) into :ypix separated by ' '
  from (select unique sticker from rawdata);
quit; run;

goptions device=png xpixels=600 ypixels=&ypix;
```

Using this technique to control the size of the page, you can use this same code to generate sparkline tables containing as few, or as many, stocks as needed. Here is an example of some output produced using the same code to plot 21 stocks:

Closing Stock Prices for Last 60 Trading Days

		07jul2011		29sep2011		low	high
Microsoft	MSFT	26.77		25.45	-4.9%	23.98	28.08
Intel	INTC	23.23		22.21	-4.4%	19.19	23.23
IBM	IBM	176.48		179.17	1.5%	157.54	185.21
Apple	AAPL	357.20		390.57	9.3%	353.21	413.45
Google	GOOG	546.60		527.50	-3.5%	490.92	622.52
Amazon	AMZN	216.74		222.44	2.6%	177.54	241.69
eBay	EBAY	33.33		30.67	-8.0%	26.95	34.42
Sirius XM	SIRI	2.22		1.49	-32.9%	1.49	2.33
Panera Bread	PNRA	131.96		107.68	-18.4%	96.68	131.96
Sony	SNE	27.29		19.73	-27.7%	18.66	27.29
Ford Motor Company	F	14.12		10.00	-29.2%	9.62	14.12
Netflix Inc	NFLX	292.42		113.19	-61.3%	113.19	298.73
WD-40 Company	WDFC	41.68		40.55	-2.7%	36.52	47.75
Verizon	VZ	37.42		37.15	-0.7%	33.12	37.57
Bank of America	BAC	10.92		6.35	-41.8%	6.06	10.92
British Petroleum	BP	44.54		37.01	-16.9%	35.73	46.77
General Motors	GM	31.80		20.76	-34.7%	20.24	31.80
Hewlett-Packard	HPQ	36.45		23.78	-34.8%	22.32	37.47
Time Warner	TWTC	21.48		16.93	-21.2%	16.05	21.49
Duke Energy Corp	DUK	19.09		20.13	5.4%	17.25	20.13
Progress Energy Inc	PGN	48.14		52.23	8.5%	43.21	52.23

Note: Each sparkline is independently auto-scaled.

EXAMPLE 15

Custom Waterfall Chart

Purpose: Create a waterfall chart in SAS/GRAPH, using the Annotate facility.

Waterfall charts provide a good way to visualize the cumulative effect of positive or negative data values sequentially added to a running total (such as a ledger or checkbook). A waterfall chart is similar to a line chart of the cumulative total, using the step interpolation.

As of SAS 9.2, SAS/GRAPH does not have a built-in easy way to produce waterfall charts, but as you might have guessed, you can create them fairly easily using a bit of custom programming.

Let's start with some simple data for a small company. After reading in these values, we can calculate the total for the end of the year, and add that total as an extra observation at the end of the data set. I call this final (modified) data set MODDATA. (Note: you could hardcode the final value in your data, but I prefer to calculate values like that programmatically to guarantee they are correct.)

```
data bardata;
input lablvar $ 1-50 barval;
datalines;
Beginning Value                                     10
First Quarter                                      -20
Second Quarter                                      30
Third Quarter                                       25
Fourth Quarter                                      37
End-of-Year Taxes                                  -40
End-of-Year Kickbacks                               11
;
run;

proc sql;
  create table sumdata as
  select 'Final Value' as lablvar, sum(barval) as barval
  from bardata;
quit; run;

data moddata;
  set bardata sumdata;
run;
```

Here is what the modified data looks like. Note that I now have an observation for the "Final Value":

```
Beginning Value           10
First Quarter            -20
Second Quarter            30
```

```
Third Quarter                      25
Fourth Quarter                     37
End-of-Year Taxes                 -40
End-of-Year Kickbacks              11
Final Value                        53
```

You could produce a simple bar chart, as follows. However, with the bars in the default alphabetical order, the chart does not reflect the sequential nature of the data:

```
title1 ls=1.5 "Simple GChart";

axis1 label=none minor=none order=(-50 to 100 by 25) offset=(0,0);
axis2 label=none value=(angle=90);

proc gchart data=moddata;
vbar lablvar / discrete type=sum sumvar=barval
 raxis=axis1 maxis=axis2
 autoref cref=gray55 lref=33 clipref;
run;
```

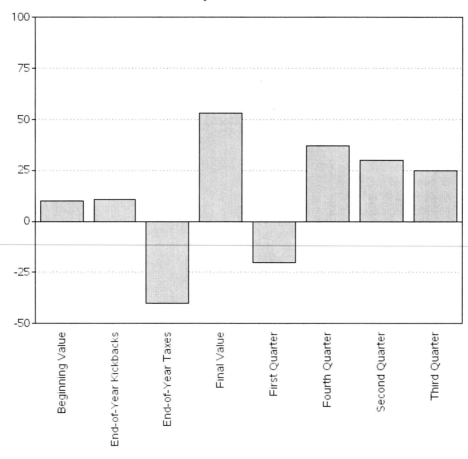

Of course you could hardcode the order of the bars in an axis statement so the bars are in the desired sequential order, as shown below. You should note, however, that hardcoding is very tedious, especially if your data changes.

```
axis2 label=none value=(angle=90)
 order=(
'Beginning Value'
```

```
'First Quarter'
'Second Quarter'
'Third Quarter'
'Fourth Quarter'
'End-of-Year Taxes'
'End-of-Year Kickbacks'
'Final Value'
  )
  ;
{re-run same GChart code as before}
```

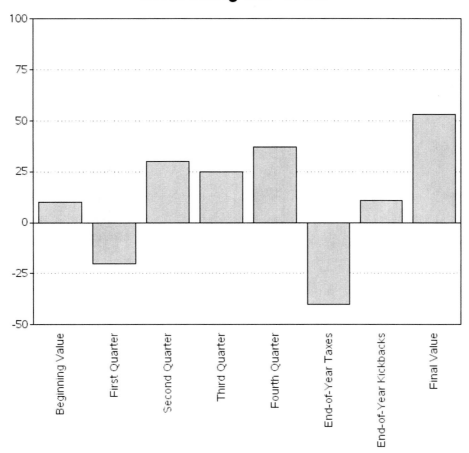

And you could even add a variable to the data, add pattern statements, and then use PROC GCHART's SUBGROUP= option to control the colors of the bars.

```
data moddata; set moddata;
 if lablvar in ('Beginning Value' 'Final Value') then colorvar=1;
 else if barval>=0 then colorvar=2;
 else colorvar=3;
run;

pattern1 v=s c=cx499DF5; /* blue */
pattern2 v=s c=cx49E20E; /* green */
pattern3 v=s c=cxFF3030; /* red */
```

```
proc gchart data=moddata;
vbar lablvar / discrete type=sum sumvar=barval
 raxis=axis1 maxis=axis2 subgroup=colorvar nolegend
 autoref cref=gray55 lref=33 clipref;
run;
```

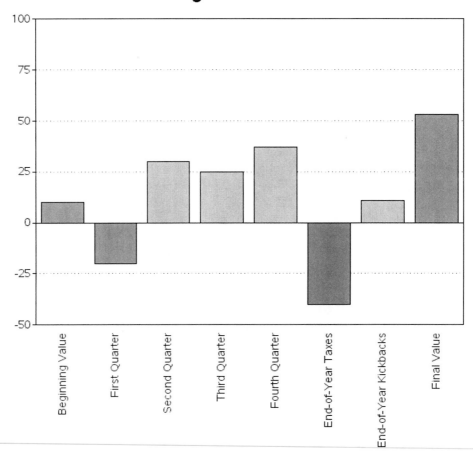

However, the result is *still* not a waterfall chart, and it does not tell you as much about the data as a waterfall chart could. For example, there is no indication of how each value affects the cumulative total, and therefore no indication whether the cumulative total dips below the zero line (which can be very important to know).

So let's get busy with some custom programming.

First, let's assign a sequential numeric value (BARNUM) to each bar, and create a user-defined format (called BAR_FMT) so those numeric values show up as the desired text values (LABLVAR). Now you can let the bars plot in their default (numeric) order, and the midpoint labels will be the text values. This technique is data-driven, eliminating the need to hardcode the text strings in the axis statement's ORDER= option (and the need to remember to hardcode it again if the data changes). Also, the numeric values make it easier to programmatically build up the Annotate commands to "draw" the waterfall bars. The code might look a little cryptic, but you do not really have to understand how it works in order to use it. (Note that this little chunk of code is very useful and very re-usable.)

```
data moddata; set moddata;
```

```
 barnum=_n_;
 fakeval=0;
run;

 data control; set moddata (rename = (barnum=start lablvar=label));
 fmtname = 'bar_fmt';
 type = 'N';
 end = START;
run;

 proc format lib=work cntlin=control;
 run;
```

You might notice that I also assigned `fakeval=0'` in the DATA step above. I use this FAKEVAL to control the bar height (`sumvar=fakeval`), and produce zero-height bars. This gives me a "blank slate" chart with all the bar midpoint labels and response axis but with no visible bars (since they are zero height).

```
 proc gchart data=moddata;
 format barnum bar_fmt.;
 vbar barnum / discrete type=sum sumvar=fakeval
  raxis=axis1 maxis=axis2 width=6
  autoref cref=gray55 lref=33 clipref
  nolegend noframe;
 run;
```

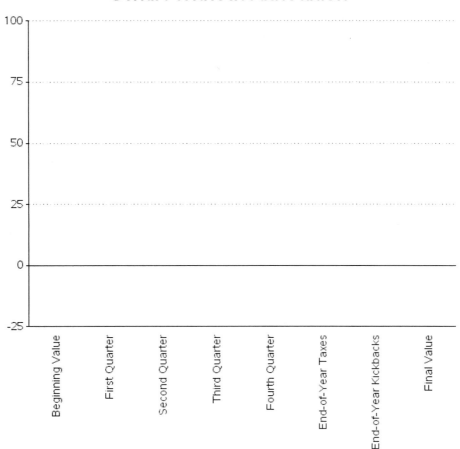

Gchart Without Annotation

The following code counts how many bars there are, so we can figure out the % width (BARWID) of each bar. The .8 value is a bit arbitrary, determined through trial-and-error, to produce visually pleasing spacing between the bars. You can adjust that value as needed.

```
proc sql noprint;
 select count(*) into :barcount from moddata;
 select (.8*(100/count(*))) into :barwid from moddata;
 select (.8*(100/count(*)))/2 into :halfwid from moddata;
quit; run;
```

The following code uses Annotate commands to draw the bars and connect them. For positive values I draw the connecting line from the top of the bar, and for negative values I draw it from the bottom—the connecting line helps you visually "follow the money." I use BY FAKEVAL merely as an easy way to let me test whether each observation is the first (first.fakeval) or last (last.fakeval) observation of the data set in order to handle those special cases.

Note that I add an HTML= variable to define the mouseover text and drill–down links (in Annotate, you use an HTML= variable, whereas in PROC GCHART you would use the HTML={some variable} option). To completely follow what each section of code is doing, see the detailed comments in the code.

```
data bar_anno; set moddata;
by fakeval;
length function $8 style $20;
when='a';
length html $500;
 html=
 'title='||quote( translate(trim(left(lablvar)),' ','\')||
    ' = '||trim(left(barval)))||
 ' href='||quote('waterfall_anno_info.htm');

/* If it's the first/left-most bar, start by moving to the zero line
*/
if first.fakeval then do;
 function='move';
 xsys='2'; x=barnum;
 ysys='2'; y=0;
 output;
 end;

/* draw a horizontal line to the midpoint of the next bar */
function='draw'; color='blue';
xsys='2'; x=barnum;
ysys='7'; y=0; /* 0% up/down, from my previous y */
output;

/* Move to the left 1/2 bar width,
   then draw a bar segment up/down based on the data +/- value */
function='move';
xsys='7'; x=-1*&halfwid; output;
function='bar'; color=barcolor; style='solid'; line=0;
xsys='7'; x=&barwid;  /* use relative-percent coordinate system */
ysys='8'; y=barval;  /* use relative-value coordinate system */
/* in the special case that it's the last bar, always connect to the
zero line instead */
if last.fakeval then do;
 ysys='2'; y=0;
 end;
```

```
output;
run;
```

Now that we have the Annotate data set, we can run the same code we used to create the blank slate chart, but this time also specify the ANNO= option, and produce a wonderful custom waterfall chart.

```
title1 ls=1.5 "SAS/GRAPH Annotated Waterfall Chart";

proc gchart data=moddata anno=bar_anno;
format barnum bar_fmt.;
vbar barnum / discrete type=sum sumvar=fakeval
 raxis=axis1 maxis=axis2 width=6
 autoref cref=gray55 lref=33 clipref
 nolegend noframe anno=anno_frame;
run;
```

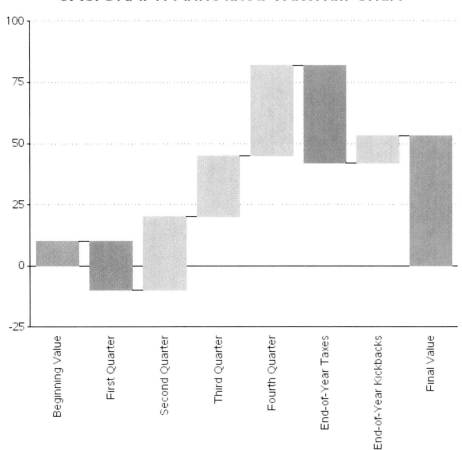

EXAMPLE 16

Plotting Data on Floor Plans

Purpose: Provide a data-driven and programmatical way to display data on an image, at the desired location.

It is often very useful to be able to show your data on a non-traditional map such as a floor plan. This example demonstrates how to use an image of your existing floor plan, and overlay your data on it using SAS/GRAPH.

You can use this technique to visualize many types of data, such as the following:

- tracking inventory or personnel on an office floor plan

- visualizing game utilization on a casino floor plan

- helping customers find items in a "big box" warehouse store

- marking the locations where artifacts were found on a site map

- displaying ride wait times on an amusement park map

- visualizing income from each seat in a sports stadium

- locating inventory in a warehouse

Years ago a technique was described in "*SAS/GRAPH Software: Examples,*"[1] for mapping locations on a floor plan. But that technique is a bit cumbersome, requiring the user to re-create all the geometry of the floor plan and rooms as a SAS/GRAPH map data set (requiring at least four X and Y coordinates per room.) Rooms that are oddly shaped or laid out at angles or along curves are very difficult to estimate X and Y coordinates for. It's a very useful technique, nevertheless, and I have used it create maps of every floor of every building at the SAS headquarters. Even so, creating all those maps was a lot of work, and the maps contain only the room outlines, showing none of the non-room information that was on the original floor plans.

So I have developed a new, simpler technique that enables you to use your floor plan image directly, and simply overlay your data onto the actual floor plan image. This technique is much less work, and produces much richer output. Here is how it works.

First, you need an image of your floor plan in a format such as JPG, PNG, GIF, or any format that the SAS/GRAPH Annotate IMAGE function can handle. In this example, I am using the floor plan map of the SUGI 30 demo floor (sugi30_floorplan.jpg). (SUGI, or SAS Users Group International, is now known as SAS Global Forum, or SGF.) Note that if you plan to make enhancements to the map at a later time, make sure that you can output the map from your floor plan design software in the exact same size, so you can use the new maps as they become available, without needing to re-estimate the locations on the map.

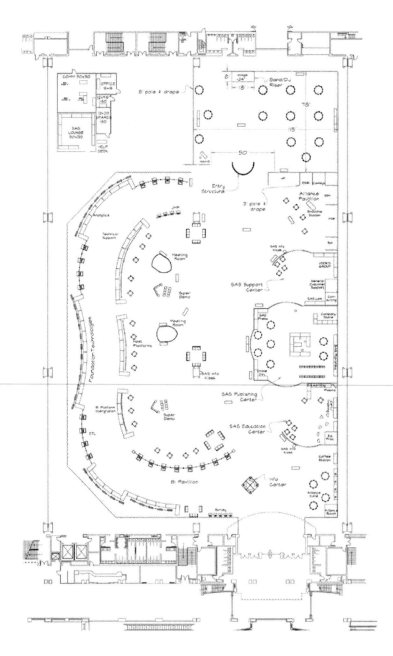

Determine the image size, in pixels, in the X and Y direction. One way to do this is to view the image in a Web browser, and right-click on the image and select **Properties**. My SUGI 30 floor plan image is 579 pixels in the X-direction, and 908 pixels in the Y-direction. I store these values as macro variables, so I can use them when needed.

```
%let xsize=579;
%let ysize=908;
```

Next, create a SAS/GRAPH map using PROC GMAP, containing a single rectangular map area that is exactly the same proportions as the floor plan image. GMAP's coordinate system starts with (0,0) at the bottom/left. The following code defines the four corners (bottom-left, bottom-right, top-right, and top-left) of the map area. The image's size (in pixels) has now become the implied coordinate system of our SAS map.

```
data floorplan_blank;
 idnum=1;
 x=0; y=0; output;
 x=&xsize; y=0; output;
 x=&xsize; y=&ysize; output;
 x=0; y=&ysize; output;
run;
```

If you want to make sure the rectangular map area is good, you can plot it with PROC GMAP. Note that no matter what size the page, PROC GMAP will always keep the map area proportional (that is, it will shrink proportionally as needed to fit on the given page).

```
proc gmap data=floorplan_blank map=floorplan_blank;
 id idnum;
 choro idnum / coutline=black;
run;
```

floorplan_blank

idnum 1

Next create an Annotate data set that will draw the floor plan image into the rectangular map area. Move to the bottom-left corner (0,0), and then fill in the image from there to the top-right corner.

```
data anno_floorplan;
 length  function $8;
```

```
xsys='2'; ysys='2'; hsys='3'; when='b';
function='move'; x=0; y=0; output;
function='image'; x=&xsize; y=&ysize; style='fit';
 imgpath="sugi30_floorplan.jpg"; output;
run;
```

Now you can annotate the image onto the map rectangle as follows:

```
pattern1 v=empty;

title1 ls=1.5 "SUGI 30 Demo Room";
proc gmap data=floorplan_blank map=floorplan_blank
anno=anno_floorplan;
 id idnum;
 choro idnum / nolegend coutline=black;
run;
```

SUGI 30 Demo Room

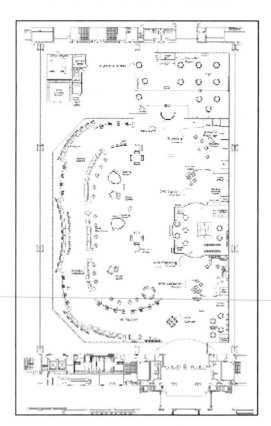

Now comes the somewhat tedious and time-consuming part: estimating an X and Y coordinate for each location in the map where you want to plot data.

In order to determine the X/Y coordinates, using the X and Y pixels of the image as the coordinate system, you can bring up the image in most any image editing software, mouseover (or click on) a certain location in the image, and see the X and Y pixel coordinates. I typically use xv on UNIX, or Paint on Microsoft Windows.

Below is a screen capture of the sugi30_floorplan.jpg in the Paint image editing software, with the pointer hovering over demo table #1, and the X/Y coordinates (266, 238) displaying in the bottom edge of the window (circled, for emphasis).

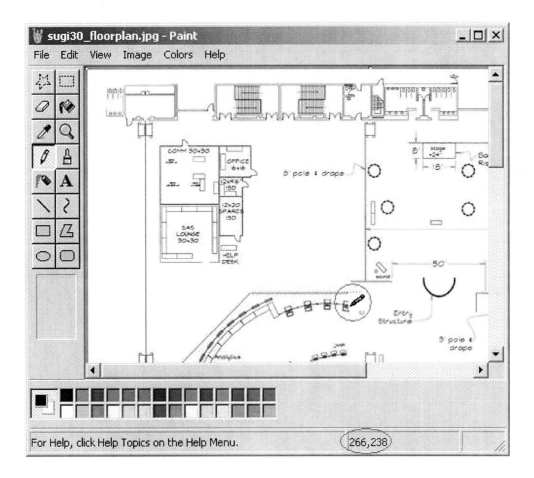

You will need to get such an estimated X and Y coordinate for each location in the map where you want to be able to plot data. In this case the demo room has numbered tables uniquely identified by a TABLE_NUM. In other maps, it might be a room number, seat, shelf, and so on. I manually look up the X and Y coordinates for each TABLE_NUM, and enter this information into a SAS data set. One caveat: most image editing software starts its coordinate system at the top left, whereas GMAP starts at the bottom left. Therefore, you need to "flip" the Y-value. You can easily do this programmatically, using y=&ysize-y. In this example, I am using the 45 demo floor tables along the left side of the SUGI 30 demo floor map. Note that the ID numbers (in my case, TABLE_NUM) need to be unique, but they do not have to be numeric.

```
data table_loc;
  input table_num x y;
  y=&ysize-y;
datalines;
1 266 238
2 246 237
3 226 239
4 205 242
{and so on}
43 235 721
44 250 723
45 260 730
;
run;
```

Now you have your map, the annotated image, and the X and Y coordinates of all the tables. All you need is some response data. In the case of the SUGI 30 demo room, I use the number of visitors per table. (Note that I am not using *real* data. These are fabricated numbers for demonstration purposes only.) In other situations your response variable might be: dollars spent, inventory count, machine uptime, and so on.

```
data attendance;
  input table_num visitors;
datalines;
1 725
2 600
3  75
4 300
{and so on}
43 825
44 750
45 800
;
run;
```

Then you merge in the X and Y coordinates of the locations, and convert the response data into an annotate data set (anno_dots). I am annotating colored dots in this example, but you could annotate dots, character font symbols, or even create a more complex glyph (where the colors and shapes tell something about the data). The possibilities are limitless.

```
proc sql;
  create table anno_dots as
  select attendance.*, table_loc.x, table_loc.y
  from attendance left join table_loc
  on attendance.table_num = table_loc.table_num;
quit; run;
```

```
data anno_dots; set anno_dots;
length function color $8;
 xsys='2'; ysys='2'; hsys='3'; when='a';
 function='pie'; rotate=360; style='psolid'; size=.55;
      if visitors<=100 then color='grayee';
 else if visitors<=200 then color='cornsilk';
 else if visitors<=300 then color='yellow';
 else if visitors<=400 then color='orange';
 else color='red';
 output;
 style='pempty'; color='gray33'; output;
run;
```

One thing that is optional, but that I recommend, is to create your output using ods html, and include an HTML variable in your Annotate data set. In that way the dots on the map have data tips; in other words, when you hover your mouse over the dots in a Web browser, you see the text data about that dot. This is useful both in validating that your map is correct, and also in letting the end user easily see more details about the data.

```
data anno_dots; set anno_dots;
length html $100;
 if style='psolid' then
   html='title='||quote('Table '||trim(left(table_num))||' had
'||trim(left(visitors))||' visitors');
run;
```

Finally, you can plot your data on the map, by specifying the ANNO= data set. Here is the code needed, including the ODS HTML statements (since the HTML data tips are so useful).

```
%let name=floorplan_dot_samples;
filename odsout '.';


GOPTIONS DEVICE=png;
ODS LISTING CLOSE;
ODS HTML path=odsout body="&name..htm" style=minimal;


goptions gunit=pct ftitle="albany amt/bold" ftext="albany amt"
htitle=3 htext=2;
goptions xpixels=&xpaper ypixels=&ypaper;


pattern1 v=empty;

title1 ls=1.5 "SUGI 30 Demo Room";

proc gmap data=floorplan_blank map=floorplan_blank
anno=anno_floorplan;
 id idnum;
 choro idnum / nolegend coutline=black anno=anno_dots
 des='' name="&name";
run;

quit;


ODS HTML CLOSE;
ODS LISTING;
```

SUGI 30 Demo Room

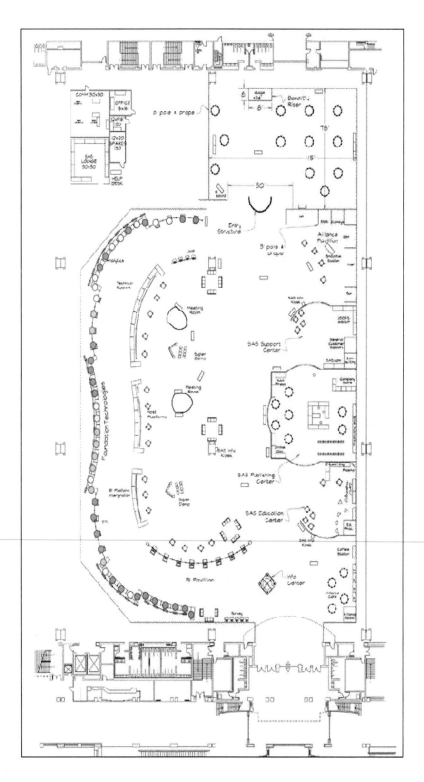

One of my favorite variations is to plot a single point for each visitor, and add a little random offset to each point's location, which produces a "density plot" of the data on the map. An advantage of this type of map is that it works better than the big colored dots in black-and-white printouts (which you will notice, if you are reading this book in black and white, as, for example, the printed version).

In the code below, the SQL is the same as before. I create an empty "invisible" pie circle (so there is an area to hold the HTML data tip). Then I run through a loop, creating a "point" for each visitor (and apply a small random offset to each X and Y coordinate).

```
proc sql;
 create table anno_density as
 select attendance.*, table_loc.x, table_loc.y
 from attendance left join table_loc
 on attendance.table_num = table_loc.table_num;
quit; run;


data anno_density; set anno_density;
 length function color $8;
 xsys='2'; ysys='2'; hsys='3'; when='b';

 function='pie'; color='white'; rotate=360; style='pempty'; size=.75;
 length html $100;
 html='title='||quote('Table '||trim(left(table_num))||' had
'||trim(left(visitors))||' visitors');
 output;

 html='';
 function='point'; color='red';
 x_center=x; y_center=y;
 do num_dots=1 to visitors;
  x=x_center+rannor(123)*4;
  y=y_center+rannor(456)*4;
  output;
 end;

run;


proc gmap data=floorplan_blank map=floorplan_blank
anno=anno_floorplan;
 id idnum;
 choro idnum / nolegend coutline=black
 anno=anno_density;
run;
```

SUGI 30 Demo Room

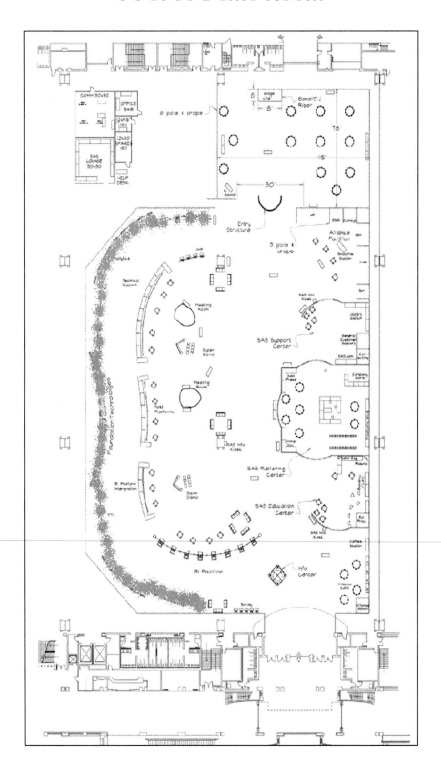

Notes

[1] SAS/Graph Software: Examples, Version 6. SAS Institute. 1993.

EXAMPLE 17

Custom Calendar Chart

Purpose: The focus of this example is a little different from most of the others. While in the other examples I am mainly trying to teach you the graphical techniques so that you can modify the techniques and create your own similar but unique graphics, in this example I am mainly demonstrating how to re-use this calendar chart with your own data.

I got the idea for this graph from Michael Friendly's Gallery of Data Visualization Web site, specifically the Bright Ideas page.[1] Below is a portion of the chart, showing four of the ten years of data.

For those of you reading this online on a color display, you can see the color in this version from Friendly's Web site:

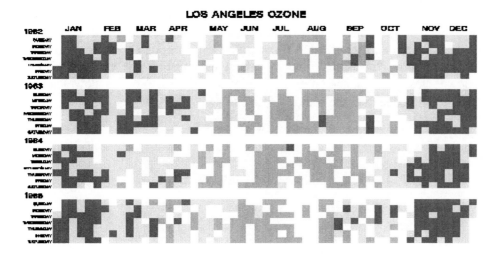

The graph also appeared in "Tracking Air Quality Trends with SAS/GRAPH," published in the SUGI 22 proceedings.[2] Here is how it appeared there, in black and white.

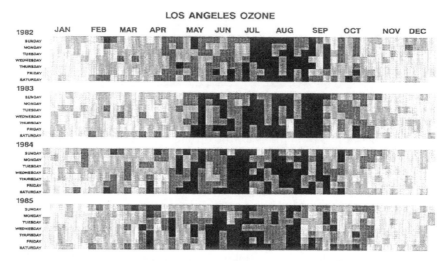

I thought this was a very interesting way to visualize data, and decided to write my own implementation using the SAS/GRAPH GMAP procedure.

The Response Data Set

To run this example, you will need some data similar to what appeared in the chart in the SUGI paper: a series of dates and values. I did not have access to the actual Los Angeles Ozone data. Therefore, I wrote a DATA step that loops through all the days in the chart, and generates some "plausible" random data. I recommend that you get the example running with this data first, and later substitute your own data. You can download both the data and the SAS code from the author's Web site.[3]

```
data my_data;
 format day date7.;
 format ozone comma5.2;
 do day = '01jan82'd to '31dec91'd by 1;
  monname=trim(left(put(day,monname.)));
  year=put(day,year.);
  ozone=ranuni(1)*.5;
  /* Simulate the ozone being better in winter months */
  if monname in ('December' 'January') then ozone=ozone/9;
   else if monname in ('November' 'February') then ozone=ozone/3.5;
   else if monname in ('September' 'October' 'March' 'April') then
ozone=ozone/1.5;
   else if ozone < .15 then ozone=.15;
  /* Simulate ozone getting better in recent years */
  if year = 1991 then ozone=.9*ozone;
  if year = 1990 then ozone=.92*ozone;
  if year = 1989 then ozone=.94*ozone;
  if year = 1988 then ozone=.95*ozone;
  if year = 1987 then ozone=.96*ozone;
  if year = 1986 then ozone=.97*ozone;
  output;
 end; run;
```

To make it easier to explain the example and demonstrate how to handle days with "missing" data, I subset the data, demonstrate with a small subset, and then later plot the entire series.

```
data my_data;
 set my_data (where=(day>='01mar1982'd and day<='15mar1982'd));
run;
```

Here is what the subset of data looks like. Each observation is just a date and a numeric value. In this example the numeric value is an ozone measurement, but you can actually plot anything you want, such as sales, attendance, temperature, stock price, number of patients, deaths, and so on.

```
  day   ozone
01MAR82 0.02
02MAR82 0.17
03MAR82 0.11
04MAR82 0.01
05MAR82 0.24
06MAR82 0.31
07MAR82 0.15
08MAR82 0.32
09MAR82 0.24
10MAR82 0.03
11MAR82 0.06
12MAR82 0.09
13MAR82 0.20
14MAR82 0.14
15MAR82 0.02
```

Now, create some macro variables based on the data values, so you will know the minimum and maximum year. While you are running PROC SQL go ahead and start your annotate data set (containing one observation for each year).

```
proc sql;
select min(year) into :min_year from my_data;
select max(year) into :max_year from my_data;
create table my_anno as select unique year from my_data;
quit; run;
```

Since your data might not have observations for every single day of the calendar (for example, the fifteen days of sample data we are working with here), loop through from January 1 of the minimum year (MIN_YEAR) to December 31 of the maximum year (MAX_YEAR), and create an observation for each day. Then merge this data set with the response data so that every day is represented in the response data.

```
data grid_days;
  format day date7.;
  do day="01jan.&min_year"d to "31dec.&max_year"d by 1;
   weekday=put(day,weekday.);
   downame=trim(left(put(day,downame.)));
   monname=trim(left(put(day,monname.)));
   year=put(day,year.);
   output;
  end;
run;
proc sql;
  create table my_data as select *
```

```
 from grid_days left join my_data
 on grid_days.day eq my_data.day;
quit; run;
```

Now, every day has a row in the response data set, either with a data value or a SAS "missing" value.

```
   day    ozone
01JAN82   .
02JAN82   .
{and so on...}
27FEB82   .
28FEB82   .
01MAR82  0.02
02MAR82  0.17
03MAR82  0.11
04MAR82  0.01
05MAR82  0.24
06MAR82  0.31
07MAR82  0.15
08MAR82  0.32
09MAR82  0.24
10MAR82  0.03
11MAR82  0.06
12MAR82  0.09
13MAR82  0.20
14MAR82  0.14
15MAR82  0.02
16MAR82   .
17MAR82   .
{and so on...}
30DEC82   .
31DEC82   .
```

There is one optional step in preparing the response data. I highly recommend creating a variable containing the HTML tags for data tips (hover text), so your users can hover their mouse over each box (day) in the calendar and see the date and data for that day. This is very useful in helping the user become oriented and see how the days are arranged in the calendar. It also provides an easy way to investigate unusual data. Use the HTML TITLE= or ALT= tag for the data tip text, and you can also include HREF= tags to add drill-down functionality. You can name this variable anything you want (I name it MYHTMLVAR in this example). When you run PROC GMAP, you just specify this variable via the HTML= option.

```
data my_data; set my_data;
 length  myhtmlvar $200;
 myhtmlvar='title='|| quote(
  put(day,downame.)||'0D'x||
  put(day,date.)||'0D'x||
  'Ozone: '||put(ozone,comma5.4)||' ppm');
run;
```

The GMAP Calendar Outline

Most SAS/GRAPH programmers think of PROC GMAP as being used only to draw geographical maps. But actually, it is a quite flexible tool that can be used to draw all sorts of things.

The way PROC GMAP works is that your MAP= data set contains a bunch of X and Y coordinates, and each group of coordinates forms a distinct area in the map (such as states in the US map). For example, below are the X and Y coordinates for the state of North Carolina from maps.us (plotted using a PROC GPLOT scatter plot, to demonstrate that they consist of a number of X/Y coordinates):

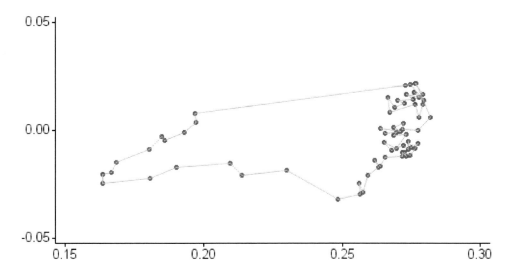

In our case, we will be creating a map where each area corresponds to a date in the calendar, with each date in the calendar being identified by 4 X and Y coordinates. Below is a PROC GPLOT scatter plot of the X and Y coordinates for January 1st, 1982:

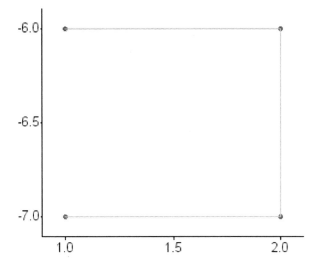

Since calendar dates all have a predictable mathematical relationship to each other, it is a fairly simple matter to come up with the X and Y coordinates for the four corners of each date in a calendar. I use a DATA step to walk my way through all the dates, and decrement Y values (that is, go down the page) based on the year and the day of the week, and increment X values (go across the page) based on the week. Basically, the calendar days are arranged top-to-bottom, left-to-right. (Note that you must sort the data first, in order to use BY YEAR, so you can test for "first.year".)

```
proc sort data=my_data out=datemap;
by year day;
run;
```

```
data datemap; set datemap;
 keep day x y;
 by year;
 if first.year then x_corner=1;
 else if trim(left(downame)) eq 'Sunday' then x_corner+1;
 y_corner=((&min_year-year)*8.5)-weekday;
 x=x_corner; y=y_corner; output;
 x=x+1; output;
 y=y-1; output;
 x=x-1; output;
run;
```

When plotted with PROC GMAP, the coordinates for January 1982 produce the following:

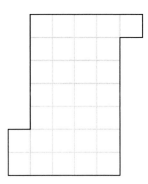

Here is the calendar map with the date annotated onto each box, so you can see how the days are arranged. This will help you see that it is like an actual calendar, flipped and turned on its side. As mentioned before, the days are arranged top-to-bottom left-to-right.

If you look at the original calendar plot, you will notice that it is difficult to determine where one month stops and another begins. I thought it would be useful to draw a darker outline around each month, to make the month boundaries easier to see (similar to a state outline around a county map). I do this using PROC GREMOVE and the Annotate facility, and the technique described in detail in Example_12, "Annotated Map Borders." The resulting Annotate data set is called "Outline."

```
data outline; set datemap;
length yr_mon $ 15;
yr_mon=trim(left(put(day,year.)))||'_'||trim(left(put(day,month.)));
order+1;
run;
```

```
proc sort data=outline out=outline;
by yr_mon order;
run;
proc gremove data=outline out=outline;
 by yr_mon; id day;
run;
data outline;
 length COLOR FUNCTION $ 8;
 retain first_x first_y;
xsys='2'; ysys='2'; size=1.75; when='A'; color='black';
 set outline; by yr_mon;
 if first.yr_mon then do;
  first_x=x; first_y=y;
  FUNCTION = 'Move'; output;
 end;
 else do;
  FUNCTION = 'Draw'; output;
 end;
 if last.yr_mon then do;
  x=first_x; y=first_y; output;
  end;
run;
```

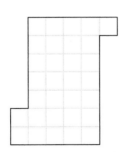

Next, I put the year and the days of the week along the left side of the chart using annotated text labels.

```
data my_anno; set my_anno;
length text $10;
function='LABEL';
position='4';
xsys='2'; ysys='2'; hsys='3'; when='A';
x=-8;
y=((&min_year-year)*8.5)-1.25;
style='';
size=2;
text=trim(left(year)); output;
x=-.1;
size=1;
text='Sunday'; output;
y=y-1; text='Monday'; output;
y=y-1; text='Tuesday'; output;
y=y-1; text='Wednesday'; output;
y=y-1; text='Thursday'; output;
y=y-1; text='Friday'; output;
y=y-1; text='Saturday'; output;
run;
```

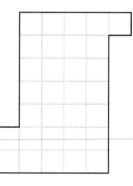

1982   Sunday
Monday
Tuesday
Wednesday
Thursday
Friday
Saturday

Then, I create an Annotate data set to place the month abbreviations along the top of the page, and combine it with the MY_ANNO_DATA set. Note that I am only showing January here (so that it easily fits on the page), but the code is also creating labels for all the months.

```
data month_anno;
length text $10;
function='LABEL';
position='5';
xsys='2'; ysys='2'; hsys='3'; when='A';
size=1.5;
y=1;
spacing=4.5;
x=(spacing/3)*-1;
x=x+spacing; text='JAN'; output;
x=x+spacing; text='FEB'; output;
x=x+spacing; text='MAR'; output;
x=x+spacing; text='APR'; output;
x=x+spacing; text='MAY'; output;
x=x+spacing; text='JUN'; output;
x=x+spacing; text='JUL'; output;
x=x+spacing; text='AUG'; output;
x=x+spacing; text='SEP'; output;
x=x+spacing; text='OCT'; output;
x=x+spacing; text='NOV'; output;
x=x+spacing; text='DEC'; output;
run;

data my_anno; set my_anno month_anno; run;
```

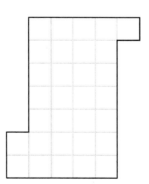

Now here is a trick to affect the size of the resulting calendar. PROC GMAP always draws the map in the largest size possible, while still preserving the X and Y aspect ratio of the map. Therefore, depending on the proportions of your page, the calendar is quite likely to extend all the way to the left (and right) edges of the page, thereby not leaving room for the annotated text. By adding an imperceptibly small "fake" map area some distance to the left of the calendar, we can guarantee there will be room for the annotated text.

```
data fake;
 day=1;
 x=-11; y=1; output;
 x=x-.001; y=y+.001; output;
 x=x+.002; output;
run;
data datemap; set datemap fake;
run;
```

Now we plot the data. Many of the options used here are optional. Note that the XPIXELS and YPIXELS need to be specified because the calendar will not fit onto the regular default page size. ODS HTML is used so the HTML mouse-over text will work. Four pattern statements are used (in conjunction with the GMAP procedure's levels=4) to control the colors of the days in the calendar. The DATA=, MAP=, and two ANNO= data sets point to the data sets created earlier.

```
%let name=calgrid;
filename odsout ".";

GOPTIONS DEVICE=png;
ODS LISTING CLOSE;
ODS HTML path=odsout body="&name..htm"
 (title="SAS/GRAPH Custom Calendar Chart Example") style=d3d;

goptions xpixels=700 ypixels=900 border cback=white;
goptions gunit=pct htitle=4 htext=2 ftitle="albany amt/bo"
ftext="albany amt";

pattern1 v=s c=cx00ff00;
pattern2 v=s c=yellow;;
pattern3 v=s c=orange;
pattern4 v=s c=red;

legend1 shape=bar(.15in,.15in) frame cshadow=gray label=none;

title "LOS ANGELES OZONE";
title2 "(using fake data)";

proc gmap data=my_data map=datemap all anno=my_anno;
 id day;
 choro ozone / levels=4
 legend=legend1 cempty=graycc
 coutline=graycc anno=outline
 html=myhtmlvar name="&name";
run;
quit;
ODS HTML CLOSE;
ODS LISTING;
```

Here is what the 15 days of sample data looks like:

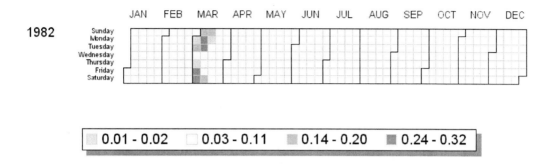

Using the data for all ten years (rather than just the fifteen days of March 1982), here is the final chart:

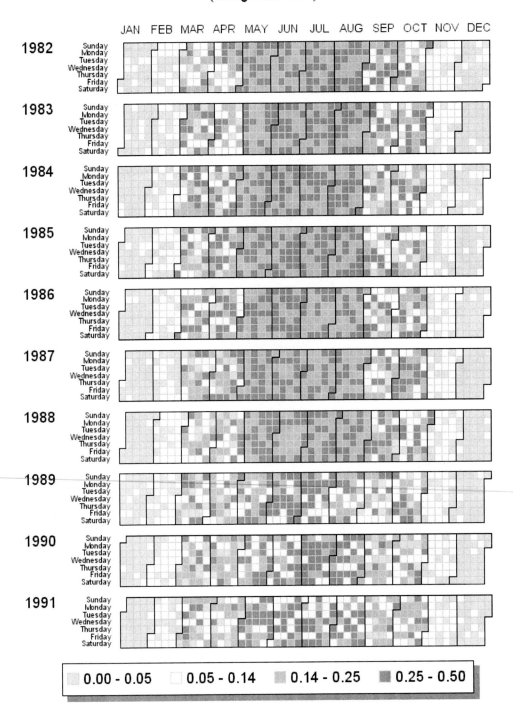

I have tried to write this example in a very generalized way, so you can easily re-use it with your own data. The only thing you should need to change is the title and the choro ozone statement.

Some final tips: If you are plotting more (or fewer) years of data, you might need to adjust the size of the page. You can do this by changing the GOPTIONS YPIXELS=, but then you will also have to change the text sizes that are scaled based on a percent of the page. These text sizes are defined in the GOPTIONS HTITLE= and HTEXT=, and also in by the SIZE= variable in the Annotate data sets. If you find yourself changing the size often, and want the annotated text to stay the same size (no matter what the YPIXELS size), then you might want to consider using hsys='D' and specifying the annotated text size in points. This is a feature added in SAS 9.2.

Also, in this example I use the simple levels=4 (quantile binning) to assign colors to the days. However, PROC GMAP supports a variety of different ways to do binning, discrete, continuous color-ramp, and so on, which you might want to learn.

Notes

[1] "Tile Maps for Temporal Patterns" example from D. Mintz, T. Fitz-Simons, and M. Wayland. 1997. http://www.datavis.ca/gallery/bright-ideas.php.
[2] "Tracking Air Quality Trends with SAS/GRAPH." *SUGI 22 Proceedings:* 807–812. Cary, NC: SAS Institute Inc.
[3] http://support.sas.com/publishing/authors/allison_robert.html.

EXAMPLE 18

Fancy Line Graph
With Broken Axis

Purpose: Demonstrate how to overlay multiple line symbols to create a "fancy" line.

I saw the original version of this graph on the biomed.brown.edu Web site, and I liked it so much that I wanted to create the same graph in SAS.

Here is the original.[1]

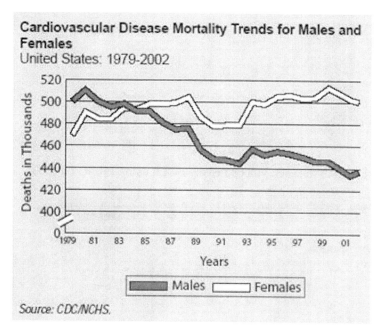

The main challenge with re-creating this graph in SAS/GRAPH is that the graph lines have a border color and SAS does not have a built-in option for such a feature. That is where a bit of custom programming, and some ingenuity, come into play.

I did not have the actual raw data, so I estimated the data values:

```
data my_data;
format date year2.;
input date date9. male_deaths female_deaths;
datalines;
01jan1979 500 470
01jan1980 510 490
01jan1981 500 483
01jan1982 496 483
01jan1983 498 492
01jan1984 490 492
01jan1985 490 498
01jan1986 480 499
01jan1987 475 500
01jan1988 478 504
01jan1989 456 482
01jan1990 448 478
01jan1991 447 479
01jan1992 444 479
01jan1993 457 500
01jan1994 452 498
01jan1995 454 503
01jan1996 451 505
01jan1997 450 503
01jan1998 446 504
01jan1999 446 513
01jan2000 440 505
01jan2001 435 502
01jan2002 438 500
;
run;
```

With the data in a SAS data set, you can easily plot it using PROC GPLOT:

```
symbol1 value=dot h=3 interpol=join w=2 color=white;
symbol2 value=dot h=3 interpol=join w=2 color=red;

proc gplot data=my_data;
  plot female_deaths*date=1 male_deaths*date=2 / overlay noframe;
run;
```

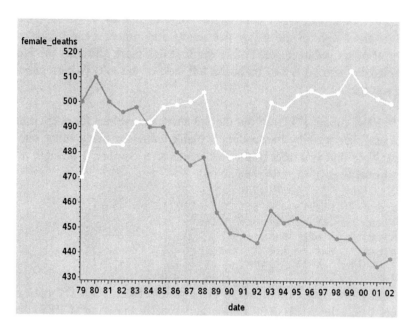

There are several standard built-in features you can use to make the graph look a lot more like the graph we are trying to imitate that don't require any custom programming.

In the code below, I add some left-justified titles at the top of the page. I could also easily include a footnote here that says "Source: CDC/NCHS" like the original graph, but since I am not using the actual data values from the CDC/NCHS, I leave that off, so as not to have anyone think my estimated values are the real values. I also add some extra white space on the left and right sides of the graph, using blank title statements angled 90 and –90 degrees (see Example 7, "Adding Border Space," for more info on that trick).

```
title1 ls=0.5 justify=left
  "  Cardiovascular Disease Mortality Trends";

title2 ls=0.5 justify=left font="albany amt/bold"
 height=4 "   for Males and Females";

title3 ls=1.0 justify=left "    United States: 1979-2002";

title4 a=90  h=2 " ";
title5 a=-90 h=6 " ";
```

You can also control the axis characteristics using simple options on axis statements, again requiring no custom programming.

On the vertical axis (`axis1`), I control the range and tick mark values using the ORDER= option, and omit minor tick marks using **minor=none**. I hardcode the axis label rather than letting it default to the name or label of the plot variable, and I use `a=90` to angle the label 90 degrees (up and down). The offset of 10 at the minimum end of the axis provides the visual space for the split axis marks I will add later.

```
axis1 order=(400 to 520 by 20) minor=none
  label=(a=90 'Deaths in Thousands') offset=(10,0);
```

On the horizontal axis (`axis2`), I control the range of tick mark values using ORDER= again. I control the height of the major tick marks and suppress the minor ones. I add a small amount of offset (white space) before the first tick mark and after the last one. I left-justify the axis label text so it will be at the left side of the axis (where people generally start reading a graph).

Here is one "tricky" thing: I "blank" out the tick mark text values for every other tick mark on the horizontal date axis by hardcoding a blank value for every other one in the axis VALUE= list. Note that you refer to the tick marks as a number, 1 through n. I blank out the even-numbered ones (`t=2`, `t=4`, and so on).

```
axis2 order=('01jan1979'd to '01jan2002'd by year)
 major=(height=.1) minor=none offset=(3,3)
 label=(justify=left 'Year')
 value=(t=2 '' t=4 '' t=6 '' t=8 '' t=10 '' t=12 ''
 t=14 '' t=16 '' t=18 '' t=20 '' t=22 '' t=24 '');
```

And here is one more easy customization via a simple PROC GPLOT option - I turn on reference lines using the AUTOREF option and specify the color (CREF) and line style (LREF). Although the original plot had solid black reference lines, I use gray dotted lines. This helps visually de-emphasize the reference lines, and keep the emphasis and attention on the data lines.

```
proc gplot data=my_data;
plot female_deaths*date=1 male_deaths*date=2 / overlay noframe
 vaxis=axis1 haxis=axis2
 autovref cvref=gray55 lvref=33;
run;
```

Here is the graph so far, with those simple changes:

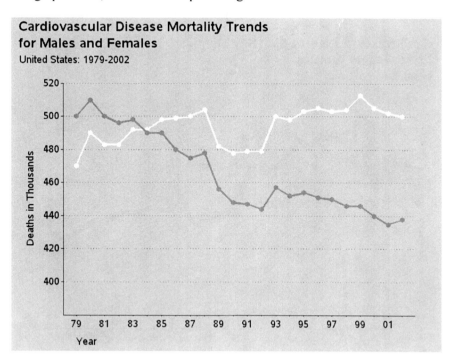

Now for the custom programming.

I often find that a single PROC GPLOT SYMBOL statement does not let me specify the exact type of line or markers I want, and I use the trick of plotting the same data multiple times, using a different symbol statement each time, to iteratively build up the complex line that I want.

In this case, I plot the data three times: first as a thick black line, then as a slightly less thick white (or red) colored line so that the thick, black line appears to outline the white line, and finally so it appears as tiny circle markers with no line. I do that last by using the W character of the MARKERE (empty marker) SAS/GRAPH software font. This produces the illusion of a thick colored line with a black outline, and the circular markers let you easily see exactly how many data points make up the line. Also if you had mouse-over text or drill-down links, these circular markers show the user where to place their mouse. Note that the order of the lines is important.

```
symbol1 i=join c=black v=none w=7;
symbol2 i=join c=white v=none w=3;
symbol3 i=none c=black f=markere v='W' h=.6;

symbol4 i=join c=black v=none w=7;
symbol5 i=join c=red v=none w=3;
symbol6 i=none c=black f=markere v='W' h=.6;

proc gplot data=my_data;
plot
 female_deaths*date=1 female_deaths*date=2 female_deaths*date=3
 male_deaths*date=4 male_deaths*date=5 male_deaths*date=6
 / overlay noframe
```

```
vaxis=axis1 haxis=axis2
autovref cvref=gray55 lvref=33;
run;
```

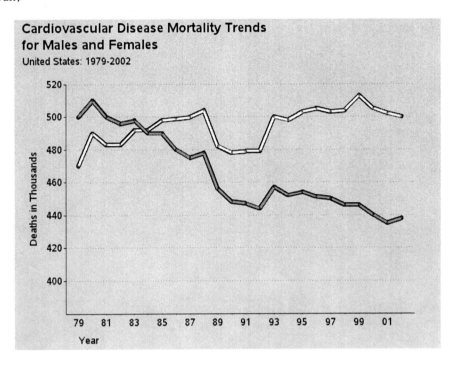

Unfortunately, when you do something tricky or custom in a graph, the automatic legends are not available, nor do they produce the desired legend. In this case, the automatic legend would show each of the six separate symbols that make up my two compound lines. But I really just want a legend showing that the red line represents males, and the white line represents females. Therefore I fake a simple legend using a footnote statement. I use the U character of the SAS/GRAPH MARKER font to draw a filled-in square, and then I move back and draw a black empty square around the filled one (using the SAS/GRAPH MARKERE empty marker font). How do you know exactly how much to move backwards, you might ask?   Trial and error!

```
footnote f=marker h=3pct c=red    'U'
         f=markere move=(-2.25,-0) c=black 'U'
         f="albany amt" h=3.5pct c=black " Males      "
         f=marker h=3pct c=white 'U'
         f=markere move=(-2.25,-0) c=black 'U'
         f="albany amt" h=3.5pct c=black " Females";
```

Now here is the final custom touch—the "break" along the vertical axis. With time series data (such as this, or stock market prices, for example), is it better to show the data plotted against a vertical axis that includes zero so the variation in the lines is visually proportional to the total quantity? Or is it better to chop the axis so that it visually magnifies the variation? Either technique is okay, but you should make it evident to the user which of the two techniques is used. In this case the "chop" technique is used; therefore, it is desirable to show a visual break in the vertical axis line, so the user does not mistakenly assume the axis starts at zero.

There is not a built-in way to do this in PROC GPLOT, but there are several ways to do it with custom programming. I chose to annotate two large / (slash) characters, angled at 33-degrees. Notice that I am using `xsys/ysys='1'` to facilitate easy placement in relation to the axis line.

```
data anno_gap;
 xsys='1'; ysys='1'; hsys='3'; position='5';
 function='label'; size=7; angle=-33;
 x=1; y=10; text='/'; output;
 x=1; y= 7; text='/'; output;
run;

proc gplot data=my_data anno=anno_gap;
 {same plot code as before}
```

Here is the final graph:

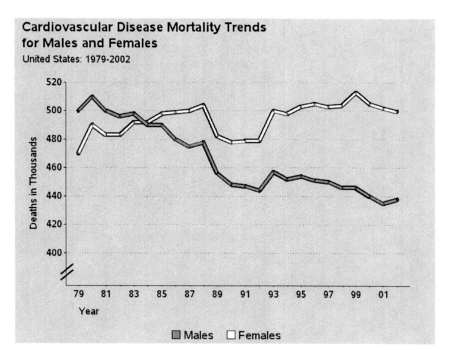

Notes

[1] From http://biomed.brown.edu/Courses/BI108/BI108_2005_Groups/02/mortality.htm. Reprinted courtesy of Albert Lin.

EXAMPLE 19

Drill-down Link to an HTML Anchor

Purpose: Create maps (or graphs) with HTML drill-down functionality, and have those drill-down links go to HTML anchors on the same page--all done in a data-driven manner.

One of the most useful features of SAS/GRAPH is the ability to create output containing HTML hotspots with data-driven drill-down functionality. When you view the output in a Web browser, you can click on these hotspots (corresponding to bars, pie slices, plot markers, or as in this case, map areas) and drill down to more detailed information.

It goes without saying that the easiest drill-down functionality to implement are links to pages that someone else maintains because you do not have to create the pages you drill down to. However, we do not usually have the luxury of someone else maintaining the *exact* drill-down pages we need, so we usually have to create our own. In the case of the U.S. map, you need to maintain 50 drill-down maps, one for each state. That is a lot of maps to maintain; therefore, you will of course want to automate that task.

This example demonstrates how to create the U.S. drill-down map, and automate the creation of the 50 individual state drill-down maps (using a macro). And as a bonus, it also demonstrates how to place all the maps in a single HTML page (rather than having a separate HTML page for each map), and have each map tagged as with an anchor that can be linked to.

Note that you can easily modify this code to work with other maps, or even other graphs; the technique is not limited to the U.S. map.

To begin, we need some data. Since the Census data is readily available, and everyone can relate to plotting population on maps, I am using year 2000 Census data from http://www.census.gov/popest/counties/files/CO-EST2004-ALLDATA.csv.

Here is some code to read in the Census CSV data file containing the population data:

```
PROC IMPORT OUT=pop_data DATAFILE="CO-EST2004-ALLDATA.csv" DBMS=CSV
REPLACE;
  GETNAMES=YES;
  DATAROW=2;
  GUESSINGROWS=1000;
RUN;
```

U.S. Map

For each state, I create a variable with an HTML data tip and drill-down link. The HTML TITLE= tag is the data tip (which displays when you move your cursor over the area in a Web browser), and the HTML HREF= tag specifies the URL to drill down to when the state is clicked.

Rather than using a normal URL, I am specifying an HTML anchor tag, which is a # followed by the state name (such as #Alaska). When you click on this link, it will jump to the #Alaska HTML anchor on your current page, which I will have added at the location of the Alaska map.

```
data pop_data; set pop_data;
length state_html $300;
state_html=
 'title='||quote(trim(left(stname))||'0d'x
         ||trim(left(put(census2000pop,comma20.0)))))||
 ' href='||quote('#'||trim(left(stname)));
run;
```

I subset just the state totals into a STATE_DATA data set, so I can plot them on the U.S. map:

```
proc sql;
 create table state_data as
 select * from pop_data
 where sumlev=40;
quit; run;
```

Now we can create the U.S. map, with hover text and drill-down links for each state, using the following code. For the HTML tags to work, you must generate the output using ODS HTML, and you must view the output using a Web browser.

```
%let name=census_map;
filename odsout '.';

ODS LISTING CLOSE;
ODS HTML path=odsout body="&name..htm";

legend1 position=(right middle) label=(position=top font="albany
amt/bold" "Population")
 shape=bar(.15in,.15in) value=(justify=center) across=1;

pattern1 v=s c=cxEDF8FB;
pattern2 v=s c=cxB3CDE3;
pattern3 v=s c=cx8C96C6;
pattern4 v=s c=cx8856A7;
pattern5 v=s c=cx810F7C;

title1 ls=1.5 "Year 2000 U.S. Census";

proc gmap data=state_data map=maps.us;
format census2000pop comma20.0;
id state;
 choro census2000pop / levels=5
 coutline=black legend=legend1
 html=state_html
 des='' name="&name";
run;
```

```
quit;
ODS HTML CLOSE;
ODS LISTING;
```

Year 2000 U.S. Census

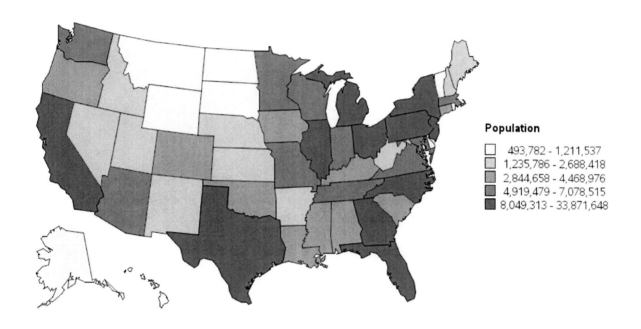

Population

☐ 493,782 - 1,211,537
▨ 1,235,786 - 2,688,418
▨ 2,844,658 - 4,468,976
▨ 4,919,479 - 7,078,515
■ 8,049,313 - 33,871,648

And now, when you view this U.S. map in a Web browser and move your cursor over a state, you will see the state name and population displayed, and when you click on the state it will drill down to the HTML anchor (on the same Web page) for that state. For example, the HTML anchor for Texas is #Texas. Note that in some browsers (such as Internet Explorer), the drill-down URL will display in the status bar along the bottom edge of the browser, as in the figure below. I have circled the anchor part of the URL so you can see it easily:

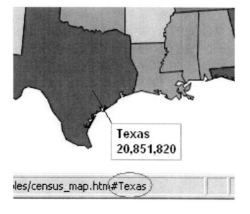

les/census_map.htm#Texas

State Maps

In order for those drill-down links to have something to actually drill down to, we need to create an individual map for each state (with an HTML anchor). The individual state maps will be shaded by the population of each county. First we will need the county data for each state, and we will set up HTML TITLE= tags with data tips showing the county name and population (but with no HREF= drill-down URLs for these individual state maps).

```
proc sql;
 create table county_data as
 select * from pop_data
 where sumlev=50;
quit; run;

data county_data; set county_data;
length county_html $300;
county_html=
 'title='||quote(trim(left(ctyname))||'0d'x
         ||trim(left(put(census2000pop,comma20.0)))));
 run;
```

To create the individual state maps, you *could* copy and paste PROC GMAP code 50 times, and edit a WHERE statement each time to create each of the 50 different state maps. That, however, is a lot of work and a lot of code to maintain.

Instead, I prefer to programmatically loop through the data and call a macro to generate each map.

First I create a data set called LOOPDATA containing one observation for each state, and then I set up a DATA step to loop through all the states and call the %DO_STATE macro once for each state (passing in the state's name to the macro). I insert this code just after the U.S. PROC GMAP code and before the ODS HTML CLOSE so that all these individual state maps are appended onto the same Web page as the U.S. map.

```
proc sql noprint;
 create table loopdata as
 select unique stname
 from state_data;
quit; run;

data _null_; set loopdata;
 call execute('%do_state('|| stname ||');');
run;
```

Of course, you will need a %DO_STATE macro, and you must declare this macro before you can use it. For that reason, I put this code at the top of the program.

The macro will need to write out an HTML anchor for the desired state (using an ODS statement)=, and then subset that state out of the U.S. county map. It will then plot that state's population data on the map. Below is a simplistic approach—a bit overly simple, but a good starting point to help you get a grasp on how everything works together.

Also note that I am using the state name (&STNAME) as part of the NAME=, so the PNG file will have a meaningful name (such as census_map_alabama.png).

```
%macro do_state(stname);

ods html anchor="&stname";

title1 ls=1.5 "Year 2000 &stname Census";

proc gmap
 data=county_data (where=(stname="&stname"))
 map=maps.uscounty (where=(fipnamel(state)="&stname")) all;
format census2000pop comma20.0;
id state county;
 choro census2000pop / levels=5
 coutline=black legend=legend1 html=county_html
 des='' name="&name._&stname";
run;

%mend do_state;
```

The above code uses the pre-projected MAPS.USCOUNTY, and simply subsets the desired state out of that map. Although this technique is simple, it produces state maps that visually look crooked. The projection (or orientation) of each state might look okay when displayed in the context of the whole U.S., but not so great when the states are displayed individually.

Year 2000 North Carolina Census

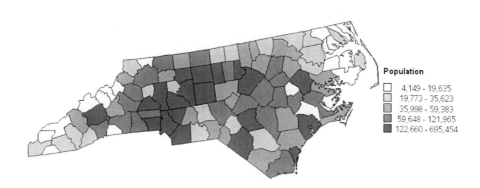

In order to make each state look good when plotted individually, you will need to project each state individually, using PROC GPROJECT. You have to use MAPS.COUNTIES (rather than MAPS.USCOUNTY) because you need unprojected latitude and longitude values to use with PROC GPROJECT. Once you have projected the map, you can plot it with PROC GMAP just like you normally would.

```
%macro do_state(stname);
ods html anchor="&stname";
```

```
proc gproject data=maps.counties (where=(fipnamel(state)="&stname"))
out=state_map
 project=robinson nodateline;
 id state county;
run;

title1 ls=1.5 "Year 2000 &stname Census";

proc gmap data=county_data (where=(stname="&stname")) map=state_map
all;
format census2000pop comma20.0;
id state county;
 choro census2000pop / levels=5
 coutline=black legend=legend1 html=county_html
 des='' name="&name._&stname";
run;

%mend do_state;
```

Here is the projected North Carolina map:

Year 2000 North Carolina Census

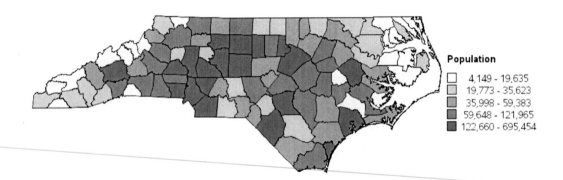

As you loop through the macro, it appends an HTML anchor, followed by the individual state map. The following shows a few anchors and states. For illustration, I have made the anchors visible so that you can see what is going on. The HTML anchors are not visible in the SAS HTML output when the actual output is viewed in a Web browser, of course.

<a name='Alabama'></a>

Year 2000 Alabama Census

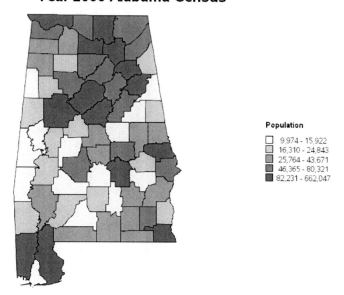

<a name='Alaska'></a>

Year 2000 Alaska Census

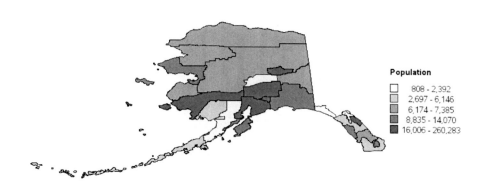

<a name='North Carolina'></a>

Year 2000 North Carolina Census

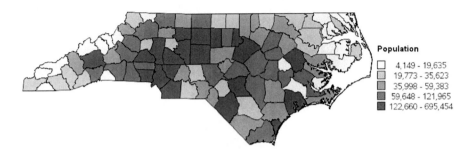

Now, when we put it all together, the SAS job first creates a U.S. map, and then appends under the U.S. map the pairs of HTML anchors and state maps for each state. When you view the HTML output, you start at the top (viewing the U.S. map) and when you click on a state in the U.S. map, you drill down to the HTML anchor for the selected state.

This technique is very useful with many types of data. And you do not have to use a map. You could use the same technique with bar charts, pie charts, scatter plots, or even tables.

EXAMPLE 20

Time Series "Strip Plot"

Purpose: Describe a technique for visually displaying very long time series.

First, let's get some long time series data. During a drought a couple of years ago, our local water supply lakes were having trouble keeping up with the water demand, and their levels were dangerously low. I was curious whether the levels had been that low in any previous years, and I found the historical lake level data and created this plot to answer my question.

Jordan Lake, near Raleigh, NC, is maintained by the Army Corps of Engineers. They monitor the lake level and make that data available on a Web page. You can actually read the data directly from their Web site using the SAS FILENAME statement URL access method, or you can save the data to your hard drive and read it using a traditional (simpler) FILENAME statement. I show the syntax for both techniques below:

```
/*
filename lakefile url "http://epec.saw.usace.army.mil/jormsr.txt"
 proxy='http://yourproxygateway.com:80';
*/

filename lakefile "jormsr.txt";
```

Rather than trying to parse the values directly out of the data file, I read each line into a long string, writing only the lines that match a certain pattern I found to be characteristic to the data lines.

```
data lakedata;
 infile lakefile firstobs=26 pad;
 input wholeline $ 1-100;
 if substr(wholeline,1,1) in ('1' '2') then output;
run;
```

I then use the SAS SCAN function to parse the values of interest out of each line of data. There are probably a dozen other ways to read and parse the data. Instead of a usual date value, the data contains a year and month. I could create a date value and hardcode the particular day as the 1st of the month. However, because SAS date values confuse some people, I chose to keep this one simple and leave the values as strictly numeric, combining the year and month (considering a month as 1/12 of a year) and storing the result in year_var.

```
data lakedata; set lakedata;
 year=.; year=scan(wholeline,1,' ');
 month=.; month=scan(wholeline,2,' ');
 year_var=year+((month-1) * 1/12);
 lake_level=.; lake_level=scan(wholeline,3,' ');
run;
```

And now I can easily plot the lake elevation over time using the following SAS/GRAPH PROC GPLOT:

```
symbol1 v=dot h=1.0 i=join c=blue;

proc gplot data=lakedata;
 plot lake_level*year_var;
run;
```

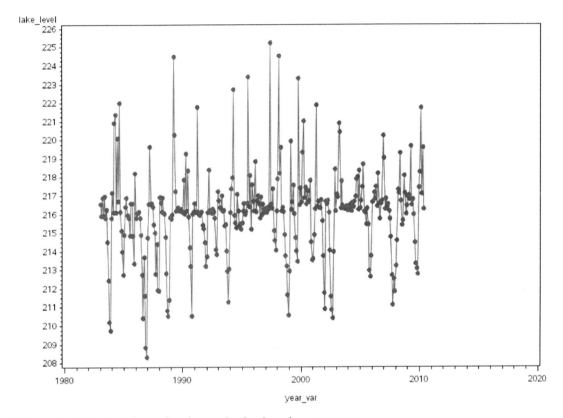

Using the default settings, the above plot is okay, but not great.

Therefore, let's make a few improvements by customizing some settings. On the axes, let's set the Y axis so that it is centered around the nominal lake level (216), and set the minimum and maximum of the X axis so that there is no wasted or blank space to the left and right of the plot line. I also add REFLINES at the beginning of each year and at the nominal 216 lake level. This makes it easy to determine if the values are above or below normal. Finally, a title is always useful to tell what the graph is about.

The default SAS/GRAPH DOT marker is not a smooth-edged TrueType-font character, but you can use any character of any font on your system for plot markers. I show two possible font markers below—one using the Windows Webdings font, and one using a character from the Cumberland AMT font that ships with SAS. I like the Webdings dot a little better, but the Cumberland AMT dot has the advantage that it can be used on any platform (Windows, UNIX, or mainframe).

```
axis1 label=(f="albany amt/bold" 'Elevation')
 order=(206 to 226 by 5) minor=none offset=(0,0);

axis2 label=(j=l f="albany amt/bold" 'Year')
 order=(1983 to 2011 by 1) minor=none offset=(0,0);

title j=left ls=1.5
 " Monthly Water Level at Jordan Lake (1983-2010) - Time Series";

symbol1 font="webdings" v='3d'x h=1.0 i=join c=blue mode=include;

/*
symbol1 font="cumberland amt/unicode" v='25cf'x h=1.4 i=join
 c=blue mode=include;
*/

proc gplot data=lakedata;
plot lake_level*year_var /
 vaxis=axis1 haxis=axis2
 vref=216 cvref=graydd
 autohref chref=graydd;
run;
```

The results shown below are better but still only okay (in my opinion). The plot is just too crowded and still does not visually convey the true nature of the time series.

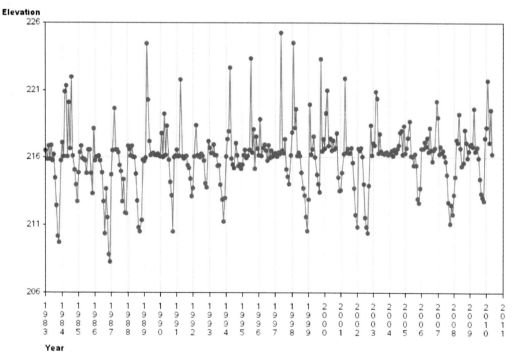

Monthly Water Level at Jordan Lake (1983-2010) - Time Series

Plot generated 29JAN11

The problem now is that the time series is just too long to be viewed in a graph of the normal proportions. To properly visualize a long time series, I believe you need to use a long plot. Luckily, SAS lets you control the horizontal and vertical proportions of the output, using the GOPTION XPIXELS= and YPIXELS= options. For long time series, I recommend setting the horizontal size (XPIXELS) to be much larger than the vertical size (YPIXELS).

```
goptions xpixels=2500 ypixels=300;

proc gplot data=lakedata;
plot lake_level*year_var /
 vaxis=axis1 haxis=axis2
 vref=216 cvref=graydd
 autohref chref=graydd;
run;
```

I have reduced the size of the following output to show the entire plot on the book page:

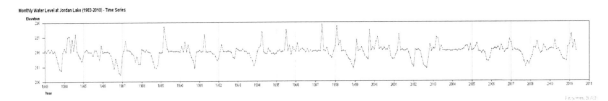

The output, however, is best viewed in a Web browser. You could view the PNG (or GIF) file directly, but Web browsers typically auto-size the output to fit in the window, and you would therefore get a tiny graph like the one above. While it can be useful to see the entire trend, the real power is in being able to scroll the output left and right. Also, you can add HTML data tips to the plot markers, so the user can move his cursor over the dots and see the actual values.

This is why I recommend using ODS HTML to create the output. The ODS HTML output forces the browser to leave the graph at full size, and enables the user to scroll left and right to see the graph. Also, ODS HTML enables you to include the HTML data tips.

```
%let name=lake_level_samples;
filename odsout '.';

ODS LISTING CLOSE;
ODS HTML path=odsout body="&name.htm" style=minimal;

/* Add html data tips to the plot markers */
data lakedata; set lakedata;
length myhtml $300;
 myhtml= 'title='||quote(
  'Level at beginning of month='||trim(left(lake_level))||'0d'x||
  'Year:'||trim(left(year))||'   month:'||
   trim(left(month))|| ' ');
run;
```

```
proc gplot data=lakedata;
plot lake_level*year_var /
 vaxis=axis1 haxis=axis2
 vref=216 cvref=graydd
 autohref chref=graydd
 html=myhtml des='' name="&name";
run;

quit;
ODS HTML CLOSE;
ODS LISTING;
```

Below is what the output looks like when viewed in a Web browser. The horizontal scroll bar enables you to see the entire time series by scrolling left and right, and you can move your cursor over the plot markers to see the data for that month. Note that the text in the HTML data tip is split into two lines—this is accomplished by including a "carriage return" (the hexadecimal character 0d) in the text.

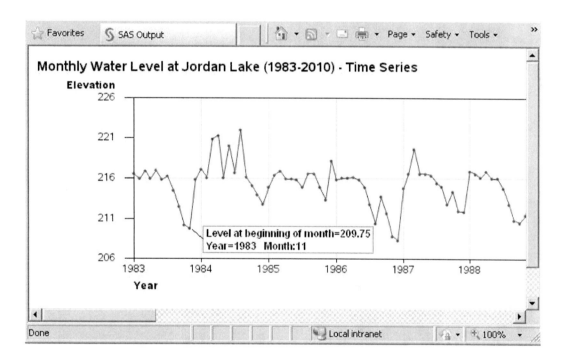

EXAMPLE 21

GIF Animation

Purpose: This example demonstrates a very powerful, yet little-known feature of SAS/GRAPH: animation.

If you want to plot some trigonometric functions, there is nothing really difficult about doing that. First, you could write a DATA step to loop through the desired angles, and calculate the value of the function or functions at each angle. For example, the following code loops from 0 to 720 degrees, and calculates the sine and cosine for each angle.

```
data mydata;
length id $10;
do Angle = 0 to 720 by 15;
  radians=Angle*(atan(1)/45);
  Value=sin(radians); id='Sine'; output;
  Value=cos(radians); id='Cosine'; output;
  end;
run;
```

And then you could use PROC GPLOT to plot the values.

```
symbol1 color=cx66CD00 v=triangle h=4.5 i=join;
symbol2 color=cx6600FF v=dot h=4.5 i=join;

title1 "GPLOT of Trig Functions";

proc gplot data=mydata;
 plot Value*Angle=id /
  autohref autovref chref=graydd cvref=graydd;
run;
```

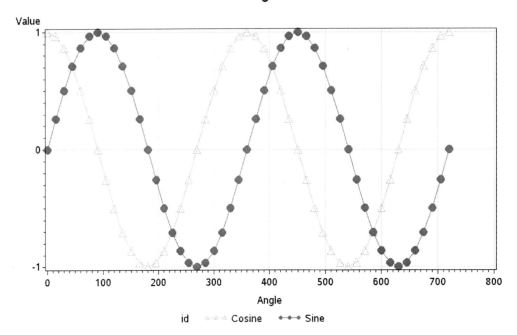

A simple plot like this can easily display the data, but sometimes it is much more powerful to use an animation, so you can see how the data changes with respect to a time variable (or in this case an angle variable).

This example demonstrates how to create animations in SAS/GRAPH. You should be able to easily adapt this code to any graphs you have with a suitable time (angle, or other) variable.

First, I will show you a very trivial example so you can see the overall process of creating a GIF animation.

Creating a GIF animation in SAS is not particularly difficult, but since there is nothing built into SAS Enterprise Guide (EG) or the SAS Output Delivery System (ODS) to help you in the process, most users do not even know it is possible. You should be able to re-use the following pieces of code to generate GIF animations from any SAS/GRAPH procedures that support DEVICE=GIF.

This first bit of code uses PUT statements to write out html code to conveniently display the GIF animation in an HTML page. This is usually handled by ODS, but ODS does not support DEV=GIFANIM.

```
%let name=wavey;
data _null_;
   file "&name..htm";
   put '<HTML>';
   put '<title>Trig Waves (SAS/Graph gifanim) </title>';
   put '<BODY>';
   put '<BLOCKQUOTE>';
   put '<P><img src="' "&name..gif" '" title="Trig Waves (SAS/Graph
gifanim)">';
   put '</BLOCKQUOTE>';
   put '</BODY>';
   put '</HTML>';
```

Next some GOPTIONS are set, so that SAS will write the graphs out in a GIF animation format with the desired settings. (Note that the HTML is written to a .htm file, and the GIF animation is written to a .gif file.)

Then you write all your graphs to the GIF animation file. Finally, the last bit of code terminates the GIF animation file. This is where you control certain things, such as the speed of the animation. (See the comments below.)

```
filename gifname "&name..gif";

goption reset
   device=gifanim
   gsfname=gifname
   gsfmode=replace          /* For the first graph, gsfmode=replace */
   disposal=background
   userinput                /* allow user input during animation */
   delay=20                 /* .20 seconds between images */
   iteration=2              /* loop animation 2 time (0=endless) */
;
```

Next, generate a separate graph for each "frame" of the animation. It is usually convenient to use a BY statement for this part, but you could generate each frame using a separate procedure invocation if you prefer. To keep this short explanation simple, I just use the imaginary procedure I call PROC G_WHATEVER. Later I show the real thing.

```
proc g_whatever data=foo;
by year;
run;
```

Finally, close off the GIF animation file by writing the hexadecimal character 3B to the file.

```
data _null_;
   file gifname recfm=n mod;
   put '3B'x;
run;
```

Here are several general tips when it comes to animating graphs:

- First ask yourself, "Does the animation help the target audience understand the data?"

- Make sure your target audience will be able to view a GIF animation. If they are using a Web browser, it will probably work, but if they are viewing it in a PDF file or printed journal, then a GIF animation is not for you.

- For each graph in the animation, be sure to identify the value of the changing variable. For example, put the value in a title.

- For each graph in the animation, try to keep everything (axes, legends, titles, and so on) the same, except for the thing you are trying to animate.

- Carefully choose the speed for your animation. It should be slow enough that users will have time to see what is going on, but not so slow that they will become bored and stop watching.

- In general, do not create endless loop animations. An animation consumes CPU, and you generally do not want it to loop endlessly.

Now, let's animate the trigonometic data.

I want to plot the trigonometric function as a line, and I have only a marker at the last point in the line, which of course, changes as the animation progresses.

I had to get a little tricky with the data to accomplish this. For each angle 0 to 720 degrees (which I call "Timestamp") I generate all the function values from 0 degrees to that angle. For the final observation for any given line, I output the last point again with a slightly different variation of the ID value, so it will use a different plot symbol (a marker instead of a line). When I plot the data, observations with id=`'SinLine'` or `'CosLine'` will be plotted as a line, and id=`'Sine'` or `'Cosine'` will be plotted as a marker. As you create custom graphs, you will find that you often need to manipulate the data in ways such as this in order to accomplish the desired custom effect.

```
data mydata1;
   length id $10;
   do Timestamp = 0 to 720 by 15;
     do Angle = 0 to Timestamp;
       radians=Angle*(atan(1)/45);
       Value = sin(radians); id='SinLine'; output;
       Value = cos(radians); id='CosLine'; output;
       if Angle eq Timestamp then do;
        Value = sin(radians); id='Sine'; output;
        Value = cos(radians); id='Cosine'; output;
       end;
     end;
   end;
run;
```

I then create a second copy of the data (MYDATA2), which I sort in the reverse order (descending), and combine with MYDATA1 to form the MYDATA data set. When I plot the data "by Timestamp," it will first animate up through time and then back down through time. You will not want, or need, to use this trick in all your animations, but I like the effect it produces in this particular data, animating the graph both forward and backward.

```
proc sort data=mydata1 out=mydata2;
 by descending Timestamp descending Angle;
run;

data mydata;
 set mydata1 mydata2;
run;
```

To make the code easily re-usable, I store the data set name and some other important values in macro variables. Keeping all the values you might need to change in one location helps make the code easy to maintain and modify.

```
%let dataset=mydata;
%let x_var=Angle;
%let y_var=Value;
%let id_var=id;
%let by_var=Timestamp;
```

There are several important things to take note of in the code below. Rather than use the automatic BY titles, I turn off the automatic BY line and insert the #BYVAL into TITLE2 text instead. Using TITLE2 titles instead of the automatic BY titles improves the look of the graph. In order to keep the axes the same in all the plots, I hardcode some axis statements rather than letting the axes auto-scale. Note that although it looks like I have two lines, I am using four symbol statements: two symbol statements are creating the lines and curves, and two symbol statements are creating the markers. I use TrueType characters from various Wingdings fonts to generate nice looking smooth-edged (TrueType) characters for the plot markers (this is possible in SAS 9.2 and higher). Note that I do not want the legend to look like it has four things in it. Therefore, I blank out the SINLINE and COSLINE in the VALUE= of the legend statement. And, to get the REFLINES at exactly the positions I want and with the colors I want, I hardcode some HREF= and VREF= options. All these little customizations work together to provide a better animation experience for the end user who is viewing the animation.

All of the following code basically is substituted for the PROC G_WHATEVER in the generic code that was shown earlier:

```
goptions noborder;
goptions cback=white;
goptions gunit=pct htitle=4 htext=3 ftitle="albany amt/bold"
 ftext="albany amt";

goptions xpixels=900 ypixels=600;

options nobyline;
title1 "Trig Wave Animation";
title2 "Angle = #byval(&by_var)";

axis1 order=(-1 to 1 by .5) value=(color=gray) minor=none
 offset=(0,0);

axis2 order=(0 to 720 by 90) value=(color=gray) minor=none
 offset=(0,0);

symbol1 color=cx66CD00 i=join value=none;
symbol2 color=cx66CD00 font="wingdings 3" v='70'x h=4.5 r=1;
 /* green */

symbol3 color=cx6600FF i=join value=none;
symbol4 color=cx6600FF font="wingdings 2" v='98'x h=4.5 r=1;
 /* purple */

legend1 label=none position=(right middle)
 value=(' ' 'Cosine' ' ' 'Sine') across=1;
```

```
proc gplot data=&dataset;
by &by_var notsorted;
plot &y_var*&x_var=&id_var /
 vref=(-.5 0 .5)
 cvref=(graydd black graydd)
 href=(90 180 270 360 450 540 630)
 chref=(graydd graydd graydd black graydd graydd graydd)
 vaxis=axis1
 haxis=axis2
 legend=legend1;
run;
```

Below is one graph (frame) out of the GIF animation. I invite you to download the code from the Web site, run it, and view the output in a Web browser so that you can see this neat little animation in action. Perhaps this will inspire you to animate some of your own time series and other data in situations where animation might help add insight into the way the data is changing over time or in conjunction with another variable.

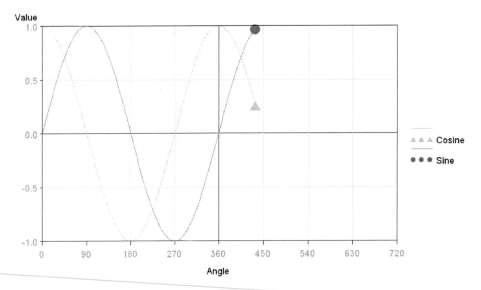

EXAMPLE 22

Using SAS/IntrNet with Graphs

Purpose: Demonstrate how to use SAS/IntrNet to generate a graph on the fly.

Sometimes it is impractical to create all the thousands of graphs a user might want ahead of time. And sometimes the user might want the graphs to contain the latest and greatest data available. Both of these scenarios can be solved by creating the graphs on the fly using SAS/IntrNet.

When a user navigates to a SAS/IntrNet URL, this triggers a SAS job to run on the server, and the graph is generated and displayed to the user in their browser. If needed, parameters can be passed on the URL, and the values of these parameters can be used in the SAS job to tell the job how to subset the data and so on.

There are entire books that cover all aspects of installing and configuring SAS/Intrnet, such as *Building Web Applications with SAS/IntrNet*, by Don Henderson. Therefore, I will not attempt to cover all those details in this short example. I will assume that you already have a SAS/IntrNet server up and running, and this example describes how to create a program that takes some user input (parameters on the URL) and generates a graph corresponding to that input.

The Biorhythm Example

No doubt, most of you have heard of biorhythms, the supposed cycles of 28, 33, and 23 days that determine your emotional, intellectual, and physical well-being. The biorhythm cycles are illustrated in the chart below. (See the color image on the Web site to discern the three colored lines.)

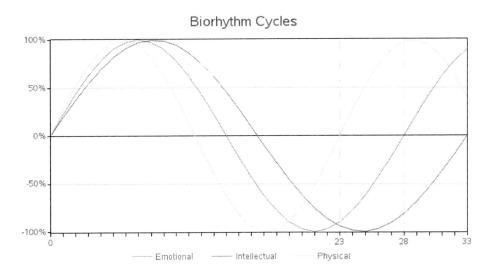

On any given day, your location on each of those three cycles can be mathematically calculated. This example demonstrates how to do that graphically. Since each user will have a different birth date, and the current day changes from day to day, this subject is a perfect candidate for a graph to be dynamically generated via SAS/IntrNet.

When SAS/IntrNet is installed, the administrator will set up a location for your SAS jobs and also a URL you can use to run those jobs. The URL will look something like the following but with your site's Web server URL and your site's values rather than appdev92 and ctntest.

```
http://sww.sas.com/sww-
bin/broker9?_service=appdev92&_program=ctntest.{job-specific stuff}
```

For this biorhythm example, my SAS job is named bio.sas, and the parameter that is passed in (and becomes a macro variable in the SAS job) is called BIRTHDATE. The complete URL to pass in the BIRTHDATE and run the bio.sas job at my site is:

```
http://sww.sas.com/sww-
bin/broker9?_service=appdev92&_program=ctntest.bio.sas&birthdate=27nov
1964
```

Here is the basic structure I use for a SAS/IntrNet job that creates a graph. You can basically use any traditional SAS/GRAPH procedure in there, and it produces ODS HTML output (complete with HTML data tips and HTML drill-down links, if your job uses those). Assuming that the job does not have to do a lot of number crunching, the graphs are usually generated in one or two seconds (depending on the speed of the server, for example). There is very little overhead, and the graphs are generated very quickly.

```
/* subset data, and/or do calculations, based on the passed-in
parameter(s) */

ods listing close;
ods html body=_webout path=&_tmpcat (url=&_replay) rs=none
style=minimal;

/* run your SAS/GRAPH proc(s), such as Proc GPlot */

quit;
ods html close;
```

The following is the complete SAS code for this SAS/IntrNet job (bio.sas). Since the focus of this example is SAS/IntrNet, rather than PROC GPLOT, I do not go into a lot of details explaining the PROC GPLOT code, which is pretty standard.

First I set some graphics options that do such things as: make sure it uses graphical text characters (GOPTIONS characters), set device to PNG, specify the nice smooth-edged Albany AMT font (especially important if the server is on UNIX), set desired XPIXELS and YPIXELS size for the PNG file, and so on.

The important thing pertaining to SAS/IntrNet is that I take the date parameter (BDATE) that was passed in on the URL and use it to control the graph. I stuff it into a temporary data set (called FOO), and create BDATE and TODAY macro variables (containing the numeric SAS date values). I then calculate a STARTDATE and ENDDATE that are +/– 30 days from today. I loop from STARTDATE to ENDDATE, and calculate the biorhythm values for physical, emotional, and intellectual cycles on each day, and plot those values using PROC GPLOT.

```
ods listing close;
ods html body=_webout path=&_tmpcat (url=&_replay) rs=none
style=minimal;

goptions cback=white characters device=png border;
goptions gunit=pct htitle=5.5 htext=3.5 ctext=gray33 ftitle="albany
amt" ftext="albany amt";
goptions xpixels=700 ypixels=375;

/* convert the text date into a sas numeric date, and put it into a
macro variable, and also put today's date (and +/- 30 days) into macro
variables. */
data foo;
bdate="&birthdate"d;
call symput('bdate',bdate);
today=today();
call symput('today',today);
startdate=today-30;
enddate=today+30;
call symput('startdate',startdate);
call symput('enddate',enddate);
run;

/* calculate biorhythm values along the 3 lines for each date */
data mydata;
```

```
  length id $15;
  format date date9.;
  format value percentn7.0;
  d2r=(atan(1)/45);  /* degrees to radians conversion factor */
  do Date = &startdate to &enddate by 1;
   t=Date-&bdate; /* days since birth */
   Value = sin((t/23)*360*d2r); id='Physical'; output;
   Value = sin((t/28)*360*d2r); id='Emotional'; output;
   Value = sin((t/33)*360*d2r); id='Intellectual'; output;
  end;
run;

title1 ls=1.5 "Biorhythm Chart";
title2 "For someone born &birthdate";

axis1 label=none order=(-1 to 1 by .50) value=(color=gray33 h=3.5)
minor=none offset=(0,0);
axis2 label=none order=(&startdate to &enddate by 30)
value=(color=gray33 h=3.5) minor=(number=29) offset=(0,0);

symbol1 color=red i=join value=none width=1;
symbol2 color=blue i=join value=none width=1;
symbol3 color=cx00ff00 i=join value=none width=1;

legend1 label=none position=(bottom center) across=3;

proc gplot data=mydata;
plot Value*Date=id /
 vref=(-.5 0 .5) cvref=(graydd black graydd)
 href=(&today) chref=(black)
 vaxis=axis1 haxis=axis2
 legend=legend1 cframe=white;
run;

quit;
ods html close;
```

Below is a snapshot of some sample output:

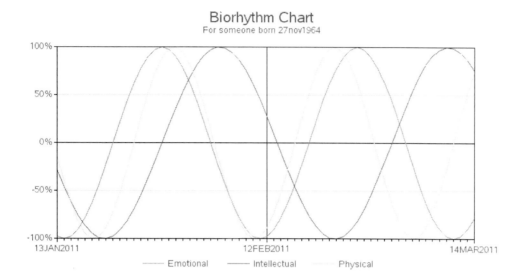

Other Possibilities for SAS/IntrNet

For the biorhythm chart, I take the passed-in date and the current date to generate the data to plot, but SAS/IntrNet has many other uses.

For example, you could pass in a value, use that value to subset your data using a WHERE clause, and then plot the results. You can also pass in multiple parameters on the URL: just separate them with ampersands (&). Or, sometimes you do not need to pass in any parameters at all. You just want to get the latest data each time the SAS/IntrNet job is run.

Also, while I did not get into it in this example, there are many ways to get the values from the user, and create the URL with those values included as the parameters. Many people use simple HTML forms or a Flash-based client. Others programmatically build the URLs in another SAS/GRAPH job, and when you click to drill down in that SAS/GRAPH output, the drill down links are the URLs that run a SAS/IntrNet job (with each bar's value as the parameter), and so on.

EXAMPLE 23

Plotting Coordinates on a Map

Purpose: Show how to plot latitude and longitude path data on a map.

As with many of my examples, this one is my imitation of an example I found on the Web. The map below originally caught my attention on Wikipedia[1]. I thought the map did a very good job of showing where tropical storms generally occur and what path they take. The map also graphically shows why the storms are called "tropical."

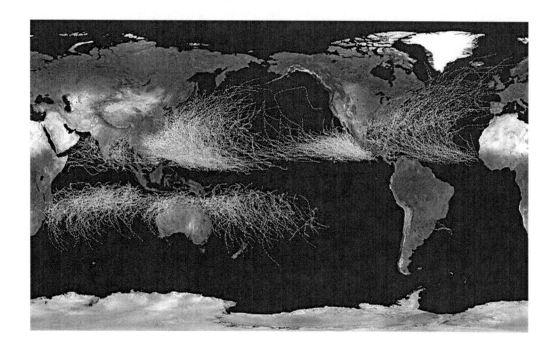

That original map is centered on the Pacific Ocean, which is a good way to map tropical storm data. However, SAS does not ship a world map arranged in that configuration. The SAS world map is centered on the Prime Meridian (LONGITUDE=0). This is why the first bit of customization is needed: in order to re-arrange the map.

I start with MAPS.WORLD from SAS, first by getting rid of the projected X and Y variables, and re-assigning the longitude and latitude to X and Y. I do this because I need X and Y to be unprojected values so that I can project them myself. While I am at it, I get rid of Antarctica (continent 97) and eliminate all the points where DENSITY is greater than 1 (so the map borders do not show too much detail).

```
data whole_map;
 set maps.world (where=(density<=1 and cont^=97) drop=x y);
 x=long;
 y=lat;
run;
```

I plot the raw unprojected latitude and longitude points using PROC GPLOT to see what the range of X values are, and thereby determine where to split the map. PROC GPLOT is a very handy tool for easily looking at map X and Y points without having to worry about map ID variables and such.

```
title1 "Gplot of lat/long radians from maps.world";
symbol1 value=dot interpol=none;
proc gplot data=whole_map;
 plot y*x / href=0;
run;
```

Gplot of lat/long radians from maps.world

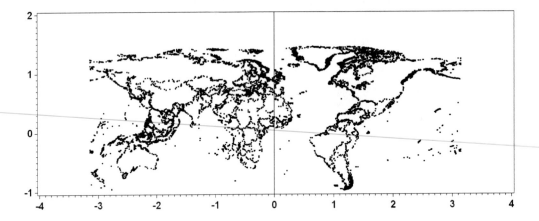

When you plot the raw X and Y points, you might notice that the land masses look backward (that is, they are a mirror image). Since this sometimes causes confusion and concern, I will try to help you understand what is going on.

When you have latitude and longitude coordinates, the longitude values start with 0 at the Prime Meridian (around London), and are usually written as the number of degrees (or radians) east or west of the Prime Meridian. For example, "80 degrees west" can be written as 80W, and "80 degrees east" as 80E. However, in a SAS data set, the longitude values are just numbers, and cannot have the "W" or "E" attached to them. Therefore, we use plus (+) and mins (–) to denote west and east, respectively. Someone had to make a decision: are values going to the west positive, or are values going to the east positive? In the case of map.world, longitude values to the west (such as the U.S.) are positive, and to the east (such as China) are denoted as negative. This is why plotting the raw (unprojected) points makes the map look backward; it is not wrong, it is just the way the data happens to be stored, and is the way PROC GPROJECT expects the X (longitude) values by default.

If you use a map where the values to the east of the Prime Meridian are positive, then you must use the EASTLONG GPROJECT option (or multiply the value by –1). The important thing to remember is that your map and Annotate X values must use the same +/– coordinate system (and sometimes you will have to multiply one or the other by –1 to make them match).

Looking at the above PROC GPLOT of the map coordinates, I determined that it would be convenient to split it at longitude value X=0. I use PROC GPROJECT with PROJECT=NONE to chop the map into two pieces. The right half contains longitude (X) values 0 to 4, and the left half contains values –4 to 0.

```
proc gproject data=whole_map out=right_map dupok project=none
  longmin=0 longmax=4;
  id id;
run;

proc gproject data=whole_map out=left_map dupok project=none
  longmin=-4 longmax=0;
  id id;
run;
```

I then apply an offset to the longitude (X) values of the right side of the map, to move it to the left side (basically, I subtract 2 pi radians). And I do a little "cheat" (by adding 5000) to make sure the segments are unique. Then I put the two halves together into one data set again.

```
data right_map; set right_map;
segment=segment+5000;
x=x-(3.14159*2);
run;

data world_map; set right_map left_map;
run;
```

Here are the re-arranged X and Y points of the map, plotted with PROC GPLOT again.

Gplot of lat/long radians from re-arranged map

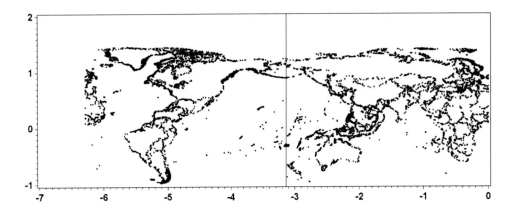

Once the latitude and longitude values are projected, the map looks correct, and we now have a world map centered on the Pacific Ocean. Here are the projected values plotted with PROC GPLOT:

Projected Points, plotted with Proc Gplot

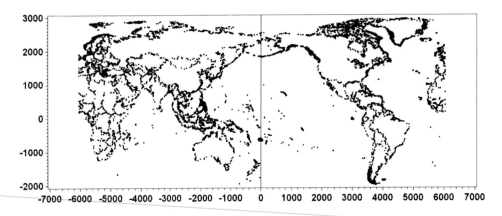

And you could now plot the map using PROC GMAP, and produce output like the following:

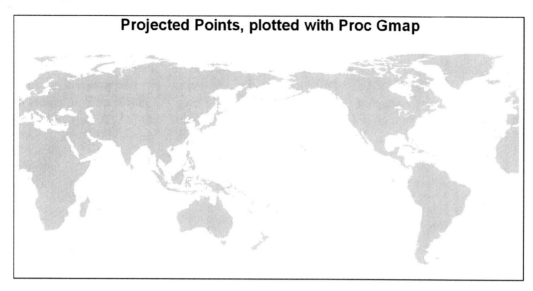

Now that we have the map, we need the hurricane data. After a bit of searching, I went to http://weather.unisys.com/hurricane/ and clicked on the "Best Track" link for each region. I then used the "Download Hurricane Database" link at the bottom of each region's page.[2]

Reading in the data was a bit tricky, but I eventually got the data into the following format (this is just the data for one storm, to keep the example simple):

```
data stormdata;
input month day time lat long wind;
region='w_pacific'; year=1959; stormnum=15; snbr=324;
datalines;
9   21      18      15.8    211.5       35
9   22      0       16.2    212.4       40
9   22      6       16.6    213.3       50
9   22      12      17.1    214.2       60
9   22      18      17.5    215.1       70
9   23      0       18.2    216.1       110
9   23      6       18.8    217.1       155
9   23      12      19.5    218.0       165
9   23      18      20.2    219.0       165
9   24      0       20.8    220.0       155
9   24      6       21.6    221.0       155
9   24      12      22.4    221.9       155
9   24      18      23.4    222.8       155
9   25      0       24.4    223.6       155
9   25      6       25.6    224.3       140
9   25      12      26.8    224.3       140
9   25      18      28.3    225.1       140
9   26      0       30.1    225.1       140
9   26      6       32.2    224.7       140
9   26      12      35.3    223.4       110
9   26      18      38.3    221.1       90
9   27      0       40.6    217.8       75
9   27      6       40.2    213.5       70
9   27      12      43.0    210.0       50
9   27      18      42.2    206.0       45
```

```
9   28      0     41.5    202.0     40
9   28      6     40.8    198.0     40
9   28     12     40.0    193.0     35
;
run;
```

I convert the longitude and latitude from degrees to radians, and apply the same –2 pi offset to the longitude (X) of the storm data as I applied to the map. I set a flag value (ANNO_FLAG) that I will use later to identify these points as Annotate (rather than map) points. And finally, I assign a variable I can use later to keep the hurricane data points in the original (chronological) order. There is a lot going on in this little DATA step.

```
data stormdata; set stormdata;
 x=atan(1)/45 * long;
 y=atan(1)/45 * lat;
 if x>0 then x=x-(3.14159*2);
 anno_flag=1;
 orig_order+1;
run;
```

Next, I combine the map and the storm data, and project the combined data set. Projecting the raw latitude and longitude coordinates enables you to control how the spherical coordinates are displayed on a flat page. There are several projection algorithms that can be used, and I chose PROJECT=CYLINDRI (cylindrical) in this example. Combining the map and the storm data and projecting them together guarantees that the projected X and Y coordinates will line up properly.

```
data combined; set world_map stormdata; run;
proc gproject data=combined out=combined dupok norangecheck
project=cylindri;
   id id;
run;
```

After projecting the combined data set, you need to separate the map and storm data into two separate data sets, so that you will have a map data set and a storm data data set. You can do this using the ANNO_FLAG variable.

```
data world_map stormdata;
   set combined;
   if anno_flag=1 then output stormdata;
   else output world_map;
run;
```

Now we start the process of converting the storm coordinates into an Annotate data set. First I find the maximum wind reading for each storm, and sort the storms so that the more powerful (that is, the more important) storms will be drawn last, and therefore will be more easily visible. Another alternative would be to sort the storms chronologically, so they will be drawn in that order. It all depends on what you want to emphasize in the map.

```
proc sql;
create table stormdata as
select unique *, max(wind) as maxwind
from stormdata
group by region, snbr;
quit; run;

proc sort data=stormdata out=stormdata;
 by maxwind region snbr orig_order;
run;
```

Next I create a variable to uniquely identify each storm. I use the combination of region and the storm number (SNBR).

```
data stormdata; set stormdata;
length byvar $20;
byvar=trim(left(region))||'_'||trim(left(snbr));
run;
```

At the beginning of each storm (FIRST.BYVAR), I lift the Annotate pen and move to that coordinate, and for each subsequent coordinate along that storm's path, I draw a line segment, thereby drawing the path of each storm.

```
data path_anno;
  length function color $8 position $1;
  retain xsys ysys '2' hsys '3' when 'a';
  set stormdata
    (keep = byvar month region snbr lat long x y wind);
  by byvar notsorted;
  position='5';
  size=.1;
  if first.byvar then do;
   function='move';
   output;
   end;
  else do;
   function='draw';
   color='red';
   output;
   end;
run;
```

Now that I have the Annotate data ready, all I have to do is draw a blank map (using PROC GMAP), and annotate the hurricane path (using `anno=path_anno`). Notice that since I am using PROC GMAP only to draw a blank map—I do not really need to specify any meaningful response data (DATA=). Therefore, I simply use the map itself as the response data, and I specify `levels=1` so that all the map areas are shaded with the color specified in PATTERN1.

```
pattern v=s c=graycc;

title1 height=.25in "Tracking Typhoon Vera in Year 1959";

goptions xpixels=1200 ypixels=600;

 proc gmap map=world_map data=world_map anno=path_anno;
 id id;
 choro id / levels=1 coutline=same nolegend;
run;
```

And there you have it: a tropical storm path plotted in the exact right location on a world map.

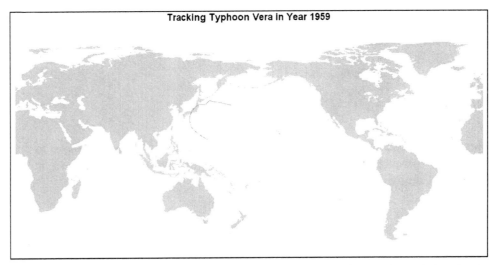

In the simplified example, I hardcode the hurricane path line to be red, but in the full-fledged example (which you can download from the author's Web page) I color-code the storms based on their wind speed.

At the top of the program, I define the colors in SAS macro variables so I can maintain them in one place and use them throughout the program.

```
%let colora=cx52b1ff;   /* lowest */
%let colorb=cx00f8f1;
%let color1=cxffffc6;
%let color2=cxffe369;
%let color3=cxffb935;
%let color4=cxff8318;
%let color5=cxff290a;   /* highest */
```

Then, when creating the Annotate data set, rather than using `color='red'`, I use the following to conditionally set the color of the line:

```
if  wind<34 then color="&colora";
else if wind<=63  then color="&colorb";
else if wind<=82  then color="&color1";
else if wind<=95  then color="&color2";
else if wind<=112 then color="&color3";
else if wind<=135 then color="&color4";
```

```
else if wind>135  then color="&color5";
else color="pink";
```

When you use the Annotate facility to create custom color graphics, you must also create your own custom legend. I sometimes use the Annotate facility to draw a legend, but in this case I use a footnote.

```
goptions cback=cx12125a;
pattern v=s c=cx5a5a5a;

title1 h=.25in c=white
  "Tracks and Intensities of Tropical Storms";

title2 h=.18in c=white "All Available Data (1851-present)";

footnote1
  f=marker h=.13in c=&colora 'U' h=.18in c=white
  f="albany amt/bold" " TD     "
  f=marker h=.13in c=&colorb 'U' h=.18in c=white
  f="albany amt/bold" " TS     "
  f=marker h=.13in c=&color1 'U' h=.18in c=white
  f="albany amt/bold" " 1      "
  f=marker h=.13in c=&color2 'U' h=.18in c=white
  f="albany amt/bold" " 2      "
  f=marker h=.13in c=&color3 'U' h=.18in c=white
  f="albany amt/bold" " 3      "
  f=marker h=.13in c=&color4 'U' h=.18in c=white
  f="albany amt/bold" " 4      "
  f=marker h=.13in c=&color5 'U' h=.18in c=white
  f="albany amt/bold" " 5"
  ;
footnote2 h=.18in c=white f="albany amt/bold"
  "Saffir-Simpson Hurricane Intensity Scale";
footnote3 h=.08 " ";
```

We use the same PROC GMAP as before, but I also specify anno=outline. This annotates the country outlines on the map (using the same technique as described in Example 12, "Annotated Map Borders"). Annotating the country outlines on top of the hurricane paths enables you to still see the country outlines, even when the country is obscured by numerous overlapping hurricanes.

```
proc gmap map=world_map data=world_map anno=path_anno;
  id id;
  choro id / levels=1 coutline=same nolegend
    anno=outline;
run;
```

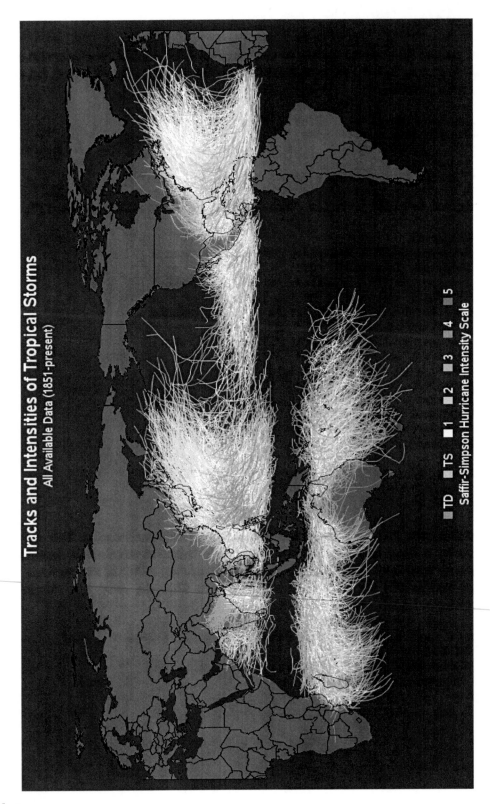

Notes

[1] http://en.wikipedia.org/wiki/File:Global_tropical_cyclone_tracks-edit2.jpg .
2 Reproduced with permission of Unisys Corporation © 2011.

EXAMPLE 24

Plotting Coverage Areas on a Map

Purpose: Create circles with a radius of a given size, and place them at the desired latitude and longitude coordinates on a map.

Jenni, one of my co-workers who volunteers with the public radio stations of North Carolina, suggested this idea for a map. She thought it would be neat to see all the North Carolina public radio stations plotted on a map, with circles showing their estimated broadcast areas. I agreed that it was an excellent idea for a custom map.

This example involves two bits of custom programming:

> Creating circles where the radius represents a geographical distance.
> Placing those circles on a map at specific latitude and longitude coordinates.

I first create a county map for the state of North Carolina (NC). I start with MAPS.COUNTIES, and subset it to just counties of NC, and just the observations where the `density<=3`. Limiting the density helps the coastline not look so cluttered and also helps PROC GPROJECT run faster (because there are fewer observations to run the calculations on). Note that you must use MAPS.COUNTIES, rather than MAPS.USCOUNTY, because MAPS.USCOUNTY does not contain the unprojected latitude and longitude values.

```
proc sql;
  create table mymap as
  select state, county, segment, x, y
  from maps.counties
  where (density<=3) and (state=stfips('NC'));
quit; run;
```

One special note about MAPS.COUNTIES: it is a bit different from most of the other maps that ship with SAS/GRAPH, in that the X and Y values are the unprojected longitude and latitude, and therefore, you can use them with PROC GPROJECT as is. By comparison, X and Y in most of the other SAS maps are projected values, and the unprojected variables called LONG and LAT must be re-named to X and Y before using those maps with PROC GPROJECT.

At this point you can use PROC GPROJECT to create the map and plot it with PROC GMAP as follows:

```
proc gproject data=mymap out=foomap;
 id state county;
run;

title1 "NC County Map";
pattern1 v=msolid c=white;
proc gmap map=foomap data=foomap;
 id state county;
 choro state / levels=1 nolegend coutline=black;
run;
```

NC County Map

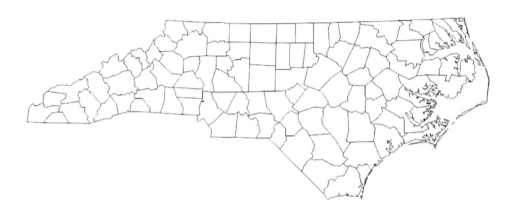

But, of course, we do not want just a map; we want a map with the radio stations plotted on it.

For that, we need some radio station location data. You can get such data from various sources on the Web, such as radio-locator.com. After a bit of copying and pasting and data wrangling, I got the data I wanted into the following format. (Note that data lines cannot be split.)

```
data stations;
input lat long other $ 19-100;
radius_miles=scan(other,4,':');
length station $20;
station=scan(other,1,':');
station=substr(station,1,index(station,'MHz')-1);
datalines;
34.75944 76.85444 WBJD-FM 91.5 MHz:C1:85000 Watts:55:Craven
    Community College, Atlantic Beach, NC
35.26722 75.53528 WBUX-FM 90.5 MHz:C2:34000 Watts:40:University of
North
    Carolina - Chapel Hill, Buxton, NC
35.58972 82.67389 WCQS-FM 88.1 MHz:C3:1600 Watts:35:Asheville, NC
35.28722 80.69583 WFAE-FM 90.7 MHz:C0:100000 Watts:75:Charlotte, NC
35.91722 80.29361 WFDD-FM 88.5 MHz:C1:60000 Watts:75:Wake Forest
    University, Winston-Salem, NC
35.84972 81.44444 WFHE-FM 90.3 MHz:C3:4000 Watts:37:Hickory, NC
35.17333 83.58111 WFQS-FM 91.3 MHz:C3:265 Watts:35:Franklin, NC
```

```
35.07278 78.89083 WFSS-FM 91.9 MHz:C1:100000 Watts:50:Fayetteville
State
    University,Fayetteville, NC
34.13139 78.18806 WHQR-FM 91.3 MHz:C:100000 Watts:77:Wilmington, NC
35.41694 77.81583 WKNS-FM 90.3 MHz:C2:35000 Watts:52:Craven Community
    College,Kinston, NC
35.73500 82.28639 WNCW-FM 88.7 MHz:C:17000 Watts:75:Isothermal
Community
    College,Spindale, NC
35.81111 77.74250 WRQM-FM 90.9 MHz:C2:7500 Watts:52:University of
North
    Carolina - Chapel Hill, Rocky Mount, NC
36.28194 76.21222 WRVS-FM 89.9 MHz:C2:41000 Watts:50:Elizabeth City
State
    University,Elizabeth City, NC
35.86639 79.16667 WUNC-FM 91.5 MHz:C:100000 Watts:85:University of
North
    Carolina,Chapel Hill, NC
35.90000 76.34583 WUND-FM 88.9 MHz:C0:50000 Watts:65:Manteo, NC
35.88667 82.55639 WYQS-FM 90.5 MHz:A:250 Watts:35:Mars Hill, NC
35.10889 77.10278 WZNB-FM 88.5 MHz:A:300 Watts:16:Craven Community
    College,New Bern, NC
36.49389 78.18972 WZRN-FM 90.5 MHz:A:2300 Watts:25:Norlina, NC
run;
```

First, let's do the easy part of the customization: annotating a dot at each latitude and longitude coordinate, with the station name just above the dot. This is fairly straightforward when using the Annotate PIE and LABEL functions.

Use `xsys'2'` and `ysys='2'` (that is, the same coordinate system as the map), and convert the X and Y from degrees to radians (because it must match the map, which is in radians). Then output a text label and a solid red PIE.

```
data stations; set stations;
  length text $25 color $12 function style $8;
  xsys='2'; ysys='2'; when='a';
  anno_flag=2;

  /* convert degrees to radians */
  x=atan(1)/45 * long;
  y=atan(1)/45 * lat;

  function='label';
  text=trim(left(scan(station,1,'-')))||' '||scan(station,2,' ');
  position='2';
  hsys='3'; size=2;
  color='black';
  output;

  function='pie';
  color="&circle_color";
  rotate=360;
  style='psolid';
  size=.4;
  output;
run;
```

Next, combine the Annotate and map data sets, run PROC GPROJECT the combined coordinates, split the two data sets apart again, and plot the map. If you do not combine the Annotate and map data sets, PROC GPROJECT will use different minimum and maximum latitude and longitude extents, and therefore the projected coordinates will not line up. (Note that a new SAS 9.3 enhancement will enable you to specify the MERIDIAN, PARALLEL1, or PARALLEL2, and you can then run PROC GPROJECT on the data sets separately, saving time by pre-projecting the map and projecting the Annotate data set only at execution time.)

```
data combined; set mymap stations; run;
proc gproject data=combined out=combined dupok;
 id state county;
run;
data mymap stations; set combined;
 if anno_flag=2 then output stations;
 else output mymap;
run;
```

When I plot the data notice that I use blank TITLE statements to add some extra white space on the left and right of the map. By default, PROC GMAP tries to make the map take up as much of the available area as possible, until it bumps into either the left or right sides, or the top or bottom. In this case, the map would go right up to the left or right sides, and there would not be room for the annotated label (it would run off the side of the page). The line coding a blank TITLE2 with a height of 4 percent, angled –90 degrees (so that it is along the right-hand side of the graph) provides space to comfortably fit the label. See Example 7, "Adding Border Space," for more info about using blank titles to add space around graphs.

```
title "Public Radio Stations in NC";
title2 a=-90 h=4pct " ";
title3 a=90 h=1pct " ";

pattern1 v=msolid c=&landcolor;
proc gmap map=mymap data=mymap anno=stations;
 id state county;
 choro state / levels=1 coutline=grayaa nolegend;
                              run;
```

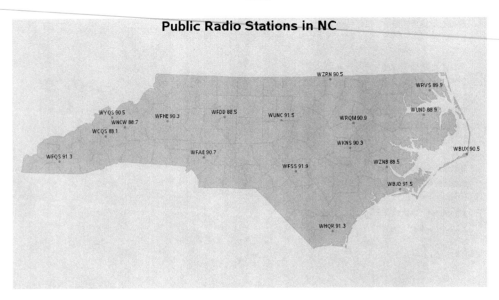

Now for the more complex part: adding the circles of a given radius. Add the following lines of code inside the same DATA step where you created the dots and labels (above).

First, define a couple of constants to use later in the equations. D2R is the "degrees to radians" conversion factor. r is the radius of the earth, in miles.

```
d2r=3.1415926/180;
r=3958.739565;
```

I then use a DO loop (inside the DATA step) to walk my way around a circle, generating the Annotate commands to draw the circle as I go. The center of the circle is at the latitude and longitude location of the radio station, and the points along the edge of the circle are calculated using geometric equations. At degree=0 you start a new polygon (function='poly') and at any other degree value you just continue the polygon (function='polycont'). This code uses a fairly simple equation that produces a reasonably good distance approximation. If you need more accuracy, you can use more complex great-circle equations, and so on.

(Note that the distance equation in this example is based on SAS Technical Support's sample #24894.[1])

```
do degree=0 to 360 by 5;

   /* Begin a new circle */
   if degree=0 then do;
      function='poly';
      style='empty';
      line=1;
      end;
   /* Continue drawing the circle */
   else do;
      function='polycont';
      color="&circle_color";
      end;

   /* Calculate x/y for the point along the circle */

y=arsin(cos(degree*d2r)*sin(radius_miles/R)*cos(lat*d2r)+cos(radius_mi
les/R)*sin(lat*d2r))/d2r;
   x=long+arsin(sin(degree*d2r)*sin(radius_miles/R)/cos(y*d2r))/d2r;
```

Finally, convert the degrees to radians, for the calculated X and Y points along the edge of the circle. Note that I am using the D2R conversion factor (because it is already easily available in a variable in this data set, from earlier calculations), whereas at other places in the code I use ATAN(1)/45. These are just two different ways of accomplishing the same thing, so do not let that confuse you.

```
   /* convert degrees to radians */
   x=d2r*x;
   y=d2r*y;
   output;
end;
```

You then plot the map using the same code as before, but this time the Annotate STATIONS data set also contains the POLY/POLYCONT Annotate instructions to draw the circle (of the desired miles radius) around each station.

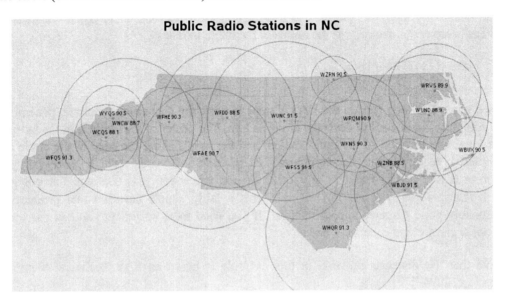

And now I will provide a little preview of a new feature that made its debut in SAS 9.3: support for transparent colors (alpha channel blending). Rather than just drawing an empty polygon for the circles, you can fill them with a transparent color. With the transparency, you will still be able to see the map through the colored circles, and the areas with multiple overlapping colored circles will show as a more concentrated (darker) color.

The following slight modification to the Annotate data set converts the previous map into the SAS 9.3 transparent version:

```
/* alpha-transparent circle areas */
data stations; set stations;
if index(function,'poly')^=0 then do;
 style='solid';
 color='aF8A10266';
 end;
run;

/* gray outlines */
data outlines; set stations (where=(function='poly' or
function='polycont'));
 style='empty';
 color='gray77';
run;

proc gmap map=mymap data=mymap anno=stations;
 id state county;
 choro state / levels=1 coutline=gray77 nolegend anno=outlines;
run;
```

This produces a map like the following:

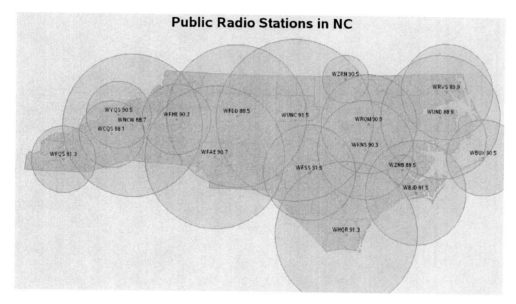

Notes

[1] "Draw data-dependent circles on a map with PROC GMAP" (http://support.sas.com/kb/24/894.html).

EXAMPLE 25

Plotting Multiple Graphs on the Same Page

Purpose: Use PROC GREPLAY to display several graphs on a page.

It is often useful to place multiple graphs on the same page, in varying sizes and positions, in order to look at the same data in different ways. There are several techniques you can use to accomplish this in SAS/GRAPH. I strongly recommend PROC GREPLAY because it provides the maximum flexibility.

To use PROC GREPLAY, you first run your code to create all the individual graphs, using the NAME= option to save them into named GRSEGs (so you can refer to them by those names later). Next create a custom PROC GREPLAY template, specifying the X and Y coordinates of the four corners of the desired areas to display each graph. And last, run PROC GREPLAY to display the saved graphs in the areas you have defined in the template.

Data

I got the original idea for this graph from an example on the DADiSP Web site: http://www.dadisp.com/grafx/XYZ-lg.gif.[1] The data can be easily calculated programmatically, using a loop in a DATA step, and calculating the values using built-in SAS trig functions.

```
data foo;
 do time = 0 to 1 by .0015;
   Angle=4*360*time;
   radians=Angle*(atan(1)/45);
   X = sin(radians);
   Y = cos(radians);
   Z = time;
   output;
 end;
run;
```

Individual Plots

Before running the code to create the plots, I set GOPTIONS NODISPLAY so the graphs are not displayed yet. For now, the GRSEGs are saved, under the NAME= names. I also set my XPIXELS andYPIXELS size so the graphs are drawn in the desired proportions (the PROC GREPLAY template and final XPIXELS andYPIXELS determine the actual size of the final output, but I always set my XPIXELS andYPIXELS to the proper sizes so that the proportions are correct).

```
goptions nodisplay;
goptions xpixels=275 ypixels=200;
goptions htitle=11pt htext=7pt ftitle="albany amt/bo" ftext="albany
amt";
```

Here is the code for the "X Data" graph (PLOT1). There is not a lot of custom programming, but there are a few tricks worth noting. One such trick is the blank TITLE2 statement, angled 90 degrees to produce extra white space to the left of the graph (and similarly a blank footnote to produce extra white space at the bottom of the graph). This white space might look strange in the individual graphs, but it produces a visually pleasing spacing in the final, combined graph. Another trick is hardcoding values for specific tick marks on the horizontal axis using `value=(t=1 '0' t=11 ' ')`. This puts a value of 0 at the first tick mark (rather than the default 0.0), and it blanks out the last tick mark value.

```
title1 ls=2.5 "X Data";
title2 a=90 h=10pct " ";
footnote1 h=10pct " ";

axis1 label=none order=(-1.5 to 1.5 by .5) minor=none offset=(0,0);
axis2 label=none order=(0 to 1 by .1) minor=none value=(t=1 '0' t=11 '
') offset=(0,0);

symbol1 i=join width=1 v=none c=red;

proc gplot data=foo;
plot x*time=1 /
 autohref chref=graydd autovref cvref=graydd
 vaxis=axis1 haxis=axis2 name="plot1";
run;
```

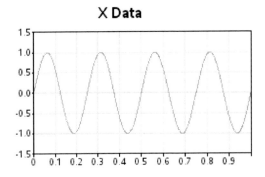

The "Y Data" graph (PLOT2) is created using almost identical code:

```
title1 ls=2.5 "Y Data";
title2 a=90 h=10pct " ";
footnote1 h=10pct " ";

symbol2 i=join width=1 v=none c=blue;

proc gplot data=foo;
plot y*time=2 /
 autohref chref=graydd autovref cvref=graydd
 vaxis=axis1 haxis=axis2 name="plot2";
run;
```

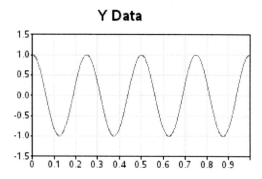

And the "Z Data" graph (PLOT3) uses a slightly different range in its vertical axis:

```
title1 ls=2.5 "Z Data";
title2 a=90 h=10pct " ";
footnote1 h=10pct " ";

axis3 label=none order=(-.2 to 1.2 by .2) minor=none offset=(0,0);

symbol3 i=join width=1 v=none c=blue;

proc gplot data=foo;
plot z*time=3 /
 autohref chref=graydd autovref cvref=graydd
 vaxis=axis3 haxis=axis2 name="plot3";
run;
```

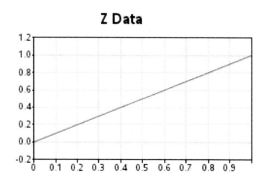

We want the three-dimensional "XYZ Data" plot (PLOT4) to be somewhat larger. Therefore, we adjust the XPIXELS andYPIXELS accordingly. I also assign a COLORVAR variable in the data to give a nice effect to the markers in the three-dimensional spiral (it helps you to visually discern which markers are at which Z depth when the spiral loops cross over one another). Note that the PROC G3D scatter plot is one of the few charts in SAS/GRAPH where you can directly assign a color via a variable in the data set (rather than having to indirectly control the color using symbol or pattern statements).

```
goptions xpixels=600 ypixels=600;

data foo; set foo;
length colorvar $8;
     if Z<=0.1 then colorvar='cx0bf5ff';
else if Z<=0.2 then colorvar='cx1de0ff';
else if Z<=0.3 then colorvar='cx3ec0ff';
else if Z<=0.4 then colorvar='cx5f9eff';
else if Z<=0.5 then colorvar='cx738aff';
else if Z<=0.6 then colorvar='cx8974ff';
else if Z<=0.7 then colorvar='cx9f5eff';
else if Z<=0.8 then colorvar='cxc33dff';
else if Z<=0.9 then colorvar='cxe21bff';
else if Z<=1.0 then colorvar='cxf607ff';
run;

title1 ls=.75 "XYZ Plot";
title2;
footnote;

proc g3d data=foo;
label X="X Data";
label Y="Y Data";
label Z="Z Data";
 scatter y*x=z / grid tilt=80 noneedle size=.4
  shape="balloon" color=colorvar name="plot4";
run;
```

XYZ Plot

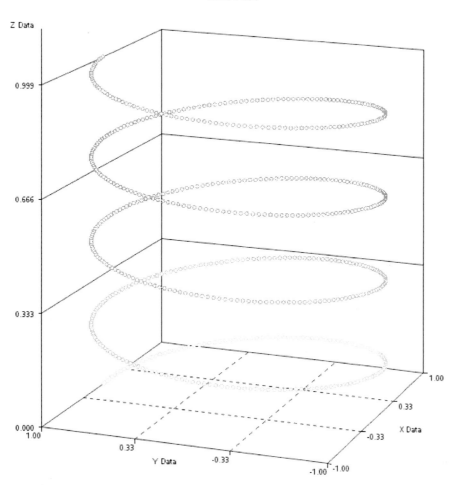

Combining the Plots

Now that we have all the individual plots, we get them all arranged on the same page using PROC GREPLAY.

I usually sketch a rough layout by hand, and then add up the total XPIXELs and YPIXELs (based on the XPIXELs and YPIXELs of the individual graphs), and calculate what percentage of the page (0–100% in the X and Y direction) each image occupies. I find it very useful to write both the pixel and percent values on the hand-drawn sketch, so I can use those numbers when I create my custom PROC GREPLAY template.

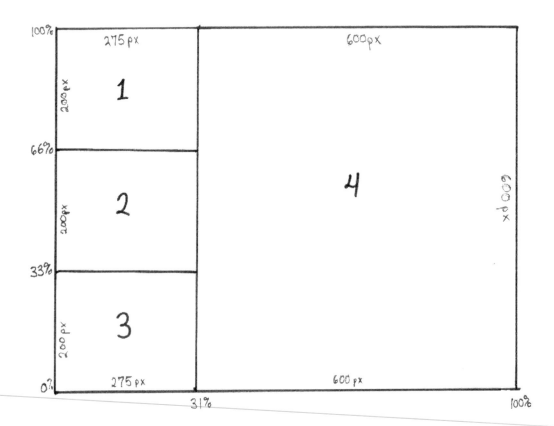

Here is the custom PROC GREPLAY template for this graph. Note that the areas are numbered 1–4 and correspond to the GRSEG names I plan to place into those areas (plot1-plot4) using PROC GREPLAY. The X and Y coordinates for the four corners of each GRSEG image are hardcoded into the template (LOWER_LEFT_X, LOWER_LEFT_Y, and so on).

```
goptions xpixels=875 ypixels=700;

proc greplay tc=tempcat nofs igout=work.gseg;
   tdef layout des='layout'

    1/llx = 0    lly = 66
      ulx = 0    uly = 100
      urx =31    ury = 100
      lrx =31    lry = 66

    2/llx = 0    lly = 33
      ulx = 0    uly = 66
      urx =31    ury = 66
      lrx =31    lry = 33

    3/llx = 0    lly = 0
      ulx = 0    uly = 33
      urx =31    ury = 33
      lrx =31    lry = 0

    4/llx =31    lly = 0
      ulx =31    uly = 100
      urx =100   ury = 100
      lrx =100   lry = 0
   ;
```

And finally, use the TREPLAY (template replay) statement to display the four graphs (plot1–plot4) into the four numbered areas (1–4) of the custom GREPLAY template. The list of template area numbers and GRSEG names are space-delimited, and the spacing does not matter to TREPLAY, but I find it useful to arrange them so they visually correspond to the page layout. In this case, 1, 2, and 3 are on the left, and 4 is on the right.

```
template = layout;
treplay
 1:plot1
 2:plot2   4:plot4
 3:plot3
 ;
run;
```

Here is the final result, with all four graphs on one page, in a visually captivating layout:

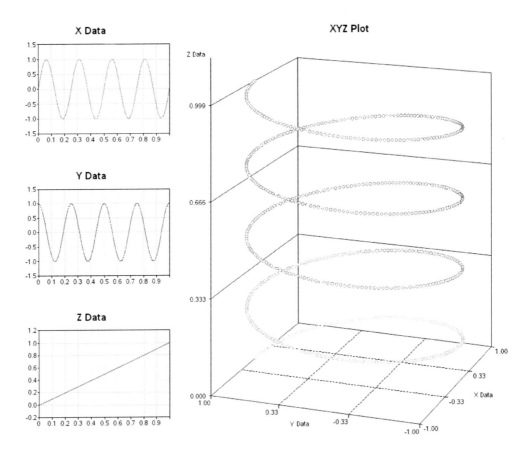

Notes

[1] From www.dadisp.com. Used by permission of DADiSP.

EXAMPLE 26

Grand Finale: An Advanced Dashboard

The final example is somewhat of a grand finale. It is a SAS/GRAPH implementation of the "Sales Dashboard" from page 177 of Stephen Few's *Information Dashboard Design.*[1] Below is the SAS output created in this example:

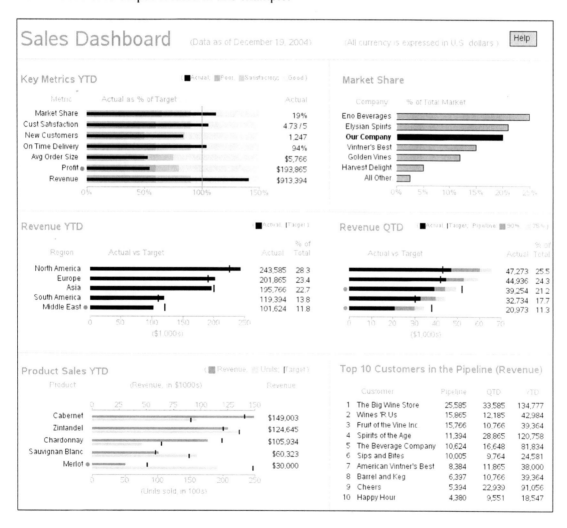

Few's book is an excellent source of information about the design of information dashboards. When he created the dashboards for his book, he used a drawing application: the advantage was that it enabled him to design the more or less "perfect" dashboard, without being hindered by the limitations of any particular software. But the disadvantage is that users have no easy way to implement dashboards exactly like the ones in the book.

I must admit that implementing Few's dashboard with SAS/GRAPH was a bit of a challenge. Even though I omitted the trend sparklines to simplify it a bit, the dashboard implementation still required almost 1000 lines of code. But the challenging nature of this dashboard also makes it a good example of what can be done with custom SAS/GRAPH programming.

Unlike the previous examples, where I provide extremely detailed descriptions of the code and show special plots that demonstrate how each of the sections of code affects the plot, I explain this example in a more general fashion. It would probably require an entire book to describe all the code in detail.

The Layout

As should be done with most complex dashboards, I first created a rough layout to determine the size and proportions. I determined that I would need one overall area (area 0) that is 900x800 pixels, and six individual graph areas. The three on the left (areas 1, 3, and 5) are 495x240 pixels, and the three on the right (areas 2, 4, and 6) are 360x240 pixels.

Macro Variables

I initialize several values in macro variables, so I can easily maintain them in one place and use them throughout the SAS job:

```
%let browntext=cxB8860B;
%let graytext=gray99;
%let div_gray=graydd;
%let backcolor=white;
goptions cback=&backcolor;

%let poorcolor=grayaa;
%let satiscolor=graycc;
%let goodcolor=grayee;

%let pipe90=gray88;
%let pipe75=graydd;

%let revcolor=gray88;
%let unitcolor=graydd;

%let font='albany amt';
%let fontb='albany amt/bold';
```

Plot 0 (Title Slide)

I use PROC GSLIDE and custom annotations to create the main title, dividing lines, and help button. This gives me total control. For example, by annotating the dividing lines (rather than using the automatic outlines provided by the GREPLAY template) I can make subtle adjustments to the line positioning, and do things such as eliminate the dividing line between the middle two graphs.

By using the Annotate LABEL function to write the text (rather than TITLE statements), I can control the exact placement and vertical alignment of the titles. I use the BAR function to draw the help button, and then annotate "Help" as text label on the button. The annotated help button uses an HTML variable with the HREF= tag to provide a drill-down link to a help page. I save the output of the PROC GSLIDE in a GRSEG named "titles" so that I can use it in PROC GREPLAY later.

```
data titlanno;
length function color $8 style $12 text $50 html $100;
xsys='3'; ysys='3'; hsys='3'; when='a';

/* annotate title text at top of page */
function='label'; style="&font";
y=97; x=.5; position='6'; color="&browntext"; size=4.3;
 text='Sales Dashboard'; output;
y=95.8; x=32; position='6'; color="&graytext"; size=1.7;
 text='(Data as of December 19, 2004)'; output;
y=95.8; x=90; position='4'; color="&graytext"; size=1.7;
 text='(All currency is expressed in U.S. dollars.)'; output;

/* draw lines to group and separate graphs */
color="&div_gray"; size=.1;

function='move'; y=92.0; x=0; output;
function='draw'; x=100; output;
```

```
function='move'; y=61.0; x=0; output;
function='draw'; x=100; output;

function='move'; y=32.0; x=0; output;
function='draw'; x=100; output;

function='move'; y=61.0; x=58; output;
function='draw'; y=92; output;

function='move'; y=32.0; x=58; output;
function='draw'; y=0; output;

/* annotate help button */
html='title='||quote('Help')||' '|| 'href="sfew_info.htm"';
function='move'; y=95; x=92; output;
function='bar'; y=y+3; x=x+5; style="solid"; color="graycc"; line=0;
output;
function='move'; y=95; x=92; output;
function='bar'; y=y+3; x=x+5; style="empty"; color="black"; line=0;
output;
line=.;
function='label'; y=95; x=92.8; position='3'; style="&font";

color="black"; size=1.7; text='Help'; output;
run;

goptions xpixels=900 ypixels=800;

title;
footnote;
proc gslide name="titles" anno=titlanno;
run;
```

Title Slide results:

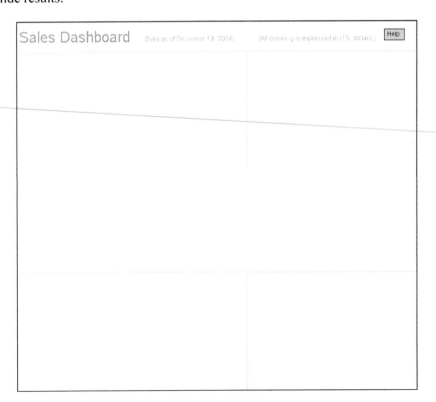

Plot 1 (Key Metrics YTD)

Data for Plot 1:

For each bar, you will need several values. The ACTUAL and TARGET values will be used to draw the bars, and the POORPCT and GOODPCT values will be used to draw the shaded areas behind the bar.

```
data data1;
input metric $ 1-20 actual target poorpct goodpct;
datalines;
Revenue              913394 650000 .60 .90
Profit               193865 360000 .60 .80
Avg Order Size         5766  11000 .50 .75
On Time Delivery        .935   .900 .60 .90
New Customers          1247   1500 .50 .85
Cust Satisfaction      4.73    4.5 .60 .90
Market Share            .190   .170 .65 .90
;
run;


/* the % of target will be the value of the black bar */
data data1; set data1;
format pct percent7.0;
pct=actual/target;
run;
```

Annotation for Plot 1:

As you might have guessed, SAS/GRAPH does not have a ready-made procedure to create this exact chart. In this case, I basically just use PROC GCHART to draw the axes and establish the coordinate system, and I annotate all the other aspects of the chart.

Of particular importance is that I annotate even the bar midpoint labels and the bars themselves, rather than using the ones that PROC GCHART draws. I annotate the midpoint labels so that I can offset them to the left, making room for the annotated red dots. And I annotate the black bars so that I can guarantee that they will be centered on top of the color ranges and will be of an appropriate size.

PROC GCHART HBAR does have an option to print a table of values to the right of the bars, but we do not actually want the bar values (which are percentages), but rather the actual values that were used behind the percent calculations. Therefore, we have to custom annotate the table. Note that I output only the labels at the top of the columns when _n_=1 (that is, the first observation of the data set). Otherwise, if you output it for every observation, you will generate multiple text labels stacked up at the same location, and the anti-aliasing will make this text appear to have a fuzzy halo.

Read the comments to see specifically what each section of Annotate code controls.

```
data anno1; set data1;
length function $8 color style $20 text $50;

/* Labels to the right of graph */
xsys='3'; ysys='2'; hsys='3'; when='a';
function='label';
x=98;
midpoint=metric;
position='4';
style="&font";
if metric in ('Revenue' 'Profit' 'Avg Order Size')
 then text=trim(left(put(actual,dollar10.0)));
else if metric in ('On Time Delivery' 'Market Share')
 then text=trim(left(put(actual,percent7.0)));
else if metric in ('New Customers')
 then text=trim(left(put(actual,comma7.0)));
else if metric in ('Cust Satisfaction')
 then text=trim(left(actual))||' / 5';
else text=trim(left(actual));
color='black';
output;

/* Red dots to left of graph */
if pct < poorpct then do;
 xsys='2'; ysys='2'; when='a';
 function='label'; color='red';
 x=0;
 midpoint=metric;
 function='move'; output;
 function='cntl2txt'; output;
 function='label'; style='"wingdings"'; position='4';
 xsys='9'; x=-.6; ysys='9'; y=+1;
 text='6c'x;
 output;
 end;

/* Labels to left of graph */
xsys='2'; ysys='2'; when='a';
function='label';
x=0;
midpoint=metric;
function='move'; output;
function='cntl2txt'; output;
function='label'; style="&font"; position='4';
xsys='9'; x=-3.0; ysys='9'; y=+1; text=metric;
color='black';
output;

/* Annotated color ranges for poor, satisfactory, and good */
when='a';  size=.;

function='move'; xsys='2'; x=0; ysys='2'; midpoint=metric; output;
 ysys='7'; y=-4; output;
function='bar'; style='solid'; line=0; color="&poorcolor"; y=+8;
x=poorpct; output;

function='move'; xsys='2'; x=poorpct; ysys='2'; midpoint=metric;
output;
 ysys='7'; y=-4; output;
```

```
function='bar'; style='solid'; line=0; color="&satiscolor"; y=+8;
x=goodpct; output;


function='move'; xsys='2'; x=goodpct; ysys='2'; midpoint=metric;
 output; ysys='7'; y=-4; output;
function='bar'; style='solid'; line=0; color="&goodcolor"; y=+8;
x=1.5; output;


line=.;


/* Now, annotate a skinnier bar to represent the data */
function='move'; xsys='2'; x=0; ysys='2'; midpoint=metric; output;
 ysys='7'; y=-1.5; output;
function='bar'; style='solid'; line=0; color="black"; y=+3; x=pct;
output;


/* Have to annotate the refline, to get it on top of the annotated
bars */
when='a';
function='move'; xsys='2'; x=1.0; ysys='1'; y=0; output;
function='draw'; line=1; color="&graytext"; y=100; size=.1; output;


/* Labels at top of the columns */
if _n_=1 then do;
 function='label'; style="&font"; color="&graytext"; size=5.5;
 ysys='1'; y=100;
 xsys='3'; x=15; position='2'; text='Metric'; output;
 xsys='1'; x=30; position='2'; text='Actual as % of Target'; output;
 xsys='3'; x=98; position='1'; text='Actual'; output;
end;


run;
```

Create Plot 1

This is basically a horizontal bar chart (PROC GCHART HBAR), with annotations
enhancing, or replacing, all aspects of the chart except for the response axis. One non-
Annotate enhancement is that I use the TITLE2 to simulate a legend because there are not
automatic legends for the color ranges we are annotating behind the bars. I save the results
in a GRSEG named PLOT1 that I can use later in PROC GREPLAY.

```
/* Use title to simulate a color legend */
title1 height=3.5 " ";
title2 height=7
 justify=left font=&fontb color=&browntext " Key Metrics YTD"
 justify=right  height=4
  font=&font      color=&graytext "( "
  font="webdings" color="black" '67'x
  font=&font      color=&graytext "Actual;   "
  font="webdings" color="&poorcolor" '67'x
  font=&font      color=&graytext "Poor;   "
  font="webdings" color="&satiscolor" '67'x
  font=&font      color=&graytext "Satisfactory;   "
  font="webdings" color="&goodcolor" '67'x
  font=&font      color=&graytext "Good )    "
 ;


/* Add white space to right, and left, of graph */
title3 angle=-90 height=25pct " ";
title4 angle=90 height=5pct " ";
```

```
axis1 label=none value=(color=&backcolor justify=right font=&font)
 order=('Revenue' 'Profit' 'Avg Order Size' 'On Time Delivery'
        'New Customers' 'Cust Satisfaction' 'Market Share')
 style=0 offset=(5,5);

axis2 label=none minor=none order=(0 to 1.5 by .5)
value=(color=&graytext)
 color=&graytext offset=(0,0);

pattern1 v=solid color=black;

goptions xpixels=495 ypixels=240;

proc gchart data=data1 anno=anno1;
 hbar metric / type=sum sumvar=pct
 width=.7 space=3
 maxis=axis1 raxis=axis2
 nostats noframe
 name="plot1";
run;
```

Plot 1 Results:

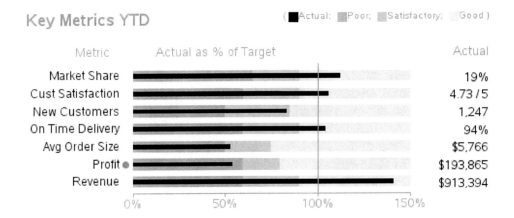

Plot 2 (Market Share)

Data for Plot 2:

The data for the second plot is a bit simpler and more straightforward. The only thing I do special in the data is setting `colorvar=2` for "Our Company."

```
data data2;
format market_share percent7.0;
input company $ 1-20 market_share;
if company eq "Our Company" then colorvar=2; else colorvar=1;
datalines;
Eno Beverages       .249
Elysian Spirits     .210
Our Company         .200
Vintner's Best      .150
Golden Vines        .120
Harvest Delight     .050
All Other           .025
;
run;
```

Annotation for Plot 2:

Similar to the first plot, we will again be annotating the bar midpoint values. This time, instead of doing this to allow space for a red dot, we annotate the labels so that we can make the "Our Company" label bold.

```
data anno2; set data2;
length function $8 color style $20 text $20;

/* Labels to left of graph */
xsys='2'; ysys='2'; hsys='3'; when='a';
function='label';
x=0;
midpoint=company;
function='move'; output;
function='cntl2txt'; output;
function='label'; style="&font"; position='4';
xsys='9'; x=-2.0; ysys='9'; y=+1; text=company;
color="black";
if company eq "Our Company" then style="&fontb";
else style="&font";
output;

/* Labels at top of the columns */
if _n_=1 then do;
 function='label'; style="&font"; color="&graytext"; size=5.5;
 ysys='1'; y=100;
 xsys='3'; x=15; position='2'; text='Company'; output;
 xsys='1'; x=30; position='2'; text='% of Total Market'; output;
end;

run;
```

Create Plot 2:

Nothing really tricky going on here; we are just running PROC GCHART HBAR and specifying the Annotate data set.

```
axis1 label=none value=(color=&backcolor justify=right font=&fontb)
 style=0 offset=(5,5);

axis2 label=none minor=none order=(0 to .25 by .05)
 value=(color=&graytext) color=&graytext offset=(0,0);

pattern1 v=solid color="&poorcolor";
pattern2 v=solid color="black";

title1 height=3.5 " ";
title2 height=7 justify=left font=&fontb color=&browntext " Market
Share";
title3 angle=-90 height=8pct " ";

goptions xpixels=360 ypixels=240;

proc gchart data=data2 anno=anno2;
hbar company / type=sum sumvar=market_share descending
subgroup=colorvar
 maxis=axis1 raxis=axis2
 width=2.0 space=1.5
```

```
   nostats noframe nolegend
   name="plot2";
run;
```

Plot 2 Results:

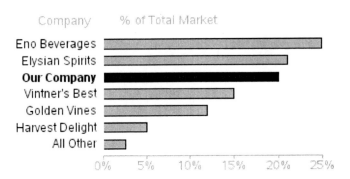

Plot 3 (Revenue YTD)

Data for Plot 3:

Plot 3's layout is similar to plot 1, but the values being plotted are very different. In this case, rather than plotting the bars against good, satisfactory, or poor color ranges, we plot the bar against a target. The values in the table to the right show the bar values as a percentage of the total (which I calculate using PROC SQL).

```
data data3;
input region $ 1-20 actual target;
actual2=actual/1000;
datalines;
North America        243585 225000
Europe               201865 190000
Asia                 195766 200000
South America        119394 110000
Middle East          101624 120000
;
run;

proc sql;
 create table data3 as
 select *,
  100*(actual/sum(actual)) as pct_total,
  (actual/target) as pct_target
 from data3;
quit; run;
```

Annotation for Plot 3:

This Annotate code is very similar to the code used in plot 1 (except the values being annotated are different). One major difference is that instead of colored areas behind the bars, I annotate a TARGET (thick line segment), using MOVE and DRAW.

```
data anno3; set data3;
length function $8 color style $20 text $20;
when='a'; size=.;

/* Annotate a bar to represent the data rather than using the gchart
bar */
function='move'; xsys='2'; x=0; ysys='2'; midpoint=region; output;
 ysys='7'; y=-3; output;
function='bar'; style='solid'; line=0; color="black"; y+6; x=actual2;
output;

/* Annotate a thick mark for the 'target' value */
function='move'; xsys='2'; x=target/1000; ysys='2'; midpoint=region;
output;
function='move'; ysys='7'; y=+6; output;
function='draw'; ysys='7'; y=-12; size=4; color="black"; output;
size=.;

/* Labels to the right of graph */
xsys='3'; ysys='2'; hsys='3'; when='a';
function='label';
midpoint=region;
position='4';
style="&font";
color='black';
x=91; text=trim(left(put(actual,comma12.0))); output;
color='black';
x=99; text=trim(left(put(pct_total,comma7.1))); output;

/* Labels to left of graph */
xsys='2'; ysys='2'; when='a';
function='label';
x=0;
midpoint=region;
function='move'; output;
function='cntl2txt'; output;
function='label'; style="&font"; position='4';
xsys='9'; x=-3.0; ysys='9'; y=+1; text=region; color='black';
color='black';
output;

/* Red dot beside 'poor' performance that needs attention */
if pct_target < .90 then do;
 xsys='2'; ysys='2'; when='a';
 function='label';
 x=0;
 midpoint=region;
 function='move'; output;
 function='cntl2txt'; output;
 function='label'; style='"wingdings"'; position='4';
 xsys='9'; x=-.6; ysys='9'; y=+1; text='6c'x; color='red'; output;
 end;
```

```
/* Annotated color ranges for poor, satisfactory, and good */
when='a';   s

/* Labels at top of the columns */
if _n_=1 then do;
 function='label'; style="&font"; color="&graytext"; size=5.5;
 ysys='1'; y=100;
 xsys='3'; x=15; position='2'; text='Region'; output;
 xsys='1'; x=30; position='2'; text='Actual vs Target'; output;
 xsys='3'; x=90; position='1'; text='Actual'; output;
 xsys='3'; x=99; position='1'; text='Total'; output;
 function='txt2cntl2'; output;
 function='move'; ysys='9'; y=1.7; output;
 function='cntl2txt'; output;
 function='label';
 xsys='3'; x=99; position='1'; text='% of'; output;
end;

run;
```

Create Plot 3:

The PROC GCHART HBAR code is almost identical to the code used for Plot 1. Note that the pink PROC GCHART BAR pattern is never seen by the end user because the annotated black bars cover the pink ones (but it is useful to be able to see the "real" bars in a different color if you omit the annotate data for debugging).

```
title1 height=3.5 " ";
title2 height=7 justify=left font=&fontb color=&browntext " Revenue
YTD"
 justify=right
  font=&font       height=4 color=&graytext "( "
  font="webdings" height=4.5 color="black" '67'x
  font=&font       height=4 color=&graytext "Actual;   "
  font=&font       height=4 color="black" '|'
  font=&font       height=4 color=&graytext "Target )   "
 ;
title3 angle=90 height=10pct " ";
title4 angle=-90 height=35pct " ";

axis1 label=none value=(color=&backcolor justify=right font=&font)
 style=0 offset=(6,6);

axis2 label=('($1,000s)') minor=none order=(0 to 250 by 50)
value=(color=&graytext)
 color=&graytext offset=(0,0);

pattern1 v=solid color=pink;

goptions xpixels=495 ypixels=240;

proc gchart data=data3 anno=anno3;
 hbar region / type=sum sumvar=actual2 descending
 width=0.8 space=2.5
 maxis=axis1 raxis=axis2
 nostats noframe
 name="plot3";
run;
```

Plot 3 Results:

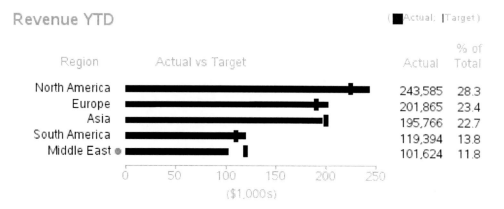

Plot 4 (Revenue QTD)

Data for Plot 4:

The data for plot 4 is very similar to the data used in plot 3, with good reason. Plots 3 and 4 share the same bar midpoint labels and therefore need to line up exactly with one another.

```
data data4;
input region $ 1-20 actual target pipe90 pipe75;
actual2=(actual/1000);
pipe90_2=(pipe90/1000);
pipe75_2=(pipe75/1000);
datalines;
North America        47273 43000 13000 5000
Europe               44936 42000  8000 6000
Asia                 39254 52000  5000 5000
South America        32734 30000  7000 4000
Middle East          20973 37500  9000 4000
;
run;

proc sql;
create table data4 as
select *,
 100*(actual/sum(actual)) as pct_total,
 (actual/target) as pct_target
from data4;
quit; run;
```

Annotation for Plot 4:

Plot 4's Annotate code is similar to the code used in previous charts. Drawing the bar segments uses code very similar to what was used to draw the color ranges behind the bars in plot 1.

```
data anno4; set data4;
length function $8 color style $20 text $20;
when='a'; size=.;
```

```
/* Annotate a bar to represent the data rather than using the gchart
bar */
function='move'; xsys='2'; x=0; ysys='2'; midpoint=region; output;
 ysys='7'; y=-3; output;
function='bar'; style='solid'; line=0; color="black"; y+6; x=actual2;
output;


/* Annotate gray bar segment for the 90% pipeline value */
function='move'; xsys='2'; x=actual2; ysys='2'; midpoint=region;
output;
 ysys='7'; y=-3; output;
function='bar'; style='solid'; line=0; color="&pipe90"; y+6;
x=actual2+pipe90_2; output;


/* Annotate gray bar segment for the 75% pipeline value */
function='move'; xsys='2'; x=actual2+pipe90_2; ysys='2';
midpoint=region; output;
 ysys='7'; y=-3; output;
function='bar'; style='solid'; line=0; color="&pipe75"; y+6;
x=actual2+pipe90_2+pipe75_2; output;


line=.;


/* Annotate a thick mark for the 'target' value */
function='move'; xsys='2'; x=target/1000; ysys='2'; midpoint=region;
output;
function='move'; ysys='7'; y=+6; output;
function='draw'; ysys='7'; y=-12; size=4; color="black"; output;
size=.;


/* Labels to the right of graph */
xsys='3'; ysys='2'; hsys='3'; when='a';
function='label';
midpoint=region;
position='4';
style="&font";
color='black';
x=89; text=trim(left(put(actual,comma12.0))); output;
x=99; text=trim(left(put(pct_total,comma7.1))); output;


/* Red dot beside 'poor' performance that needs attention */
if pct_target < .90 then do;
 xsys='2'; ysys='2'; when='a';
 function='label';
 x=0;
 midpoint=region;
 function='move'; output;
 function='cntl2txt'; output;
 function='label'; style='"wingdings"'; position='4';
 xsys='9'; x=-.6; ysys='9'; y=+1; text='6c'x; color='red'; output;
 end;


when='a';  size=.;


/* Labels at top of the columns */
if _n_=1 then do;
 function='label'; style="&font"; color="&graytext"; size=5.5;
 ysys='1'; y=100;
 xsys='1'; x=30; position='2'; text='Actual vs Target'; output;
 xsys='3'; x=89; position='1'; text='Actual'; output;
 xsys='3'; x=99; position='1'; text='Total'; output;
 function='txt2cntl2'; output;
```

```
function='move'; ysys='9'; y=1.7; output;
function='cntl2txt'; output;
function='label';
xsys='3'; x=99; position='1'; text='% of'; output;
end;

run;
```

Create Plot 4:

Once again, we use TITLE2 to simulate a legend, and then save the output in a GRSEG named PLOT4 (which we will use later in PROC GREPLAY).

```
title1 height=3.5 " ";
title2 height=7 justify=left font=&fontb color=&browntext "Revenue
QTD"
 justify=right height=4
  font=&font       color=&graytext "( "
  font="webdings" color="black" '67'x
  font=&font       color=&graytext "Actual; "
  font=&font       color="black" '|'
  font=&font       color=&graytext "Target;   Pipeline: "
  font="webdings" color="&pipe90" '67'x
  font=&font       color=&graytext " 90%;   "
  font="webdings" color="&pipe75" '67'x
  font=&font       color=&graytext " 75% )   "
 ;

title3 angle=90 height=1pct " ";
title4 angle=-90 height=27pct " ";

axis1 label=none value=(color=&backcolor justify=right font=&font
height=.5)
 style=0 offset=(6,6);

axis2 label=('($1,000s)') minor=none order=(0 to 70 by 10)
value=(color=&graytext)
 color=&graytext offset=(0,0);

pattern1 v=solid color=pink;
goptions xpixels=360 ypixels=240;
proc gchart data=data4 anno=anno4;
 hbar region / type=sum sumvar=actual2 descending
 width=0.8 space=2.5
 maxis=axis1 raxis=axis2
 nostats noframe
 name="plot4";
run;
```

Plot 4 Results:

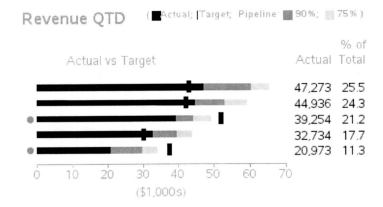

Plot 5 (Product Sales YTD)

Data for Plot 5:

Because we have to deal with grouped bars, this one is a little more difficult than the others. For example, we have to manipulate the data and generate two observations for each observation in the original data.

```
%let maxrevenue=150;
%let maxunits=250;

data data5;
input product $ 1-20 revenue rev_target units unit_target;
revenue2=revenue/1000;
units2=units/100;
rev_target2=rev_target/1000;
unit_target2=unit_target/100;
datalines;
Cabernet            149003 140000 16000 15000
Zinfandel           124645 120000 22500 22500
Chardonnay          105934 119000  7500 10500
Sauvignan Blanc      60323  58000 16000 14800
Merlot               30000  50000 19000 24600
;
run;

/* Re-arrange the data for grouped bar chart */
data data5; set data5;
groupvar=product;
barvar='Revenue'; value=revenue2; target=rev_target2;
pct_target=value/target;
 normval=value/&maxrevenue; normtarg=target/&maxrevenue;
 output;
barvar='Units';   value=units2;   target=unit_target2;
pct_target=value/target;
 normval=value/&maxunits; normtarg=target/&maxunits;
 output;
run;
```

Annotation for Plot 5:

The annotate data is also a bit more complex, because rather than dealing with MIDPOINT and X values, we also have to include a GROUP value.

```
data anno5; set data5;
length function $8 color style $20 text $50;
when='a'; size=.;

/* Annotate a bar to represent the data rather than using the gchart
bar */
if barvar eq 'Revenue' then color="&revcolor"; else
color="&unitcolor";
function='move'; xsys='2'; x=0; ysys='2'; group=groupvar;
midpoint=barvar; output;
ysys='7'; y=-1.5; output;
function='bar'; style='solid'; line=0; y=+3; x=normval; output;
line=.;

/* Annotate a thick mark for the 'target' value */
function='move'; xsys='2'; x=normtarg; ysys='2'; group=groupvar;
midpoint=barvar; output;
function='move'; ysys='7'; y=+3; output;
function='draw'; ysys='7'; y=-6; size=4; color="black"; output;
size=.;

/* Labels to the right of graph */
xsys='3'; ysys='2'; hsys='3'; when='a';
function='label';
group=groupvar;
midpoint=barvar;
position='4';
style="&font";
color='black';
if barvar eq 'Revenue' then do;
 x=95; text=trim(left(put(revenue,dollar12.0)));
 output;
 end;

/* Labels to left of graph */
xsys='2'; ysys='2'; when='a';
function='label';
x=0;
group=groupvar;
midpoint=barvar;
function='move'; output;
function='cntl2txt'; output;
function='label'; style="&font"; position='4';
xsys='9'; x=-3.0; ysys='9'; y=+1; text=groupvar; color='black';
color='black';
if barvar eq 'Revenue' then output;

/* Red dot beside 'poor' performance that needs attention */
if pct_target < .70 then do;
 xsys='2'; ysys='2'; when='a';
 function='label';
 x=0;
 group=groupvar;
 midpoint=barvar;
```

```
 function='move'; output;
 function='cntl2txt'; output;
 function='label'; style='"wingdings"'; position='4';
 xsys='9'; x=-.6; ysys='9'; y=+1; text='6c'x; color='red'; output;
 end;
run;
```

The PROC GCHART HBAR statement does not support having two axes (a separate axis for the two bars in the grouped bar pairs). Therefore, we need to annotate the axis line and the tick marks at the top of the graph.

```
data anno5b;
length function $8 color style $20 text $50;
when='a'; size=.; hsys='3';

/* Annotate axis line at top of graph */
color="&graytext"; line=1;
function='move'; xsys='1'; ysys='1'; x=0; y=100; output;
function='draw'; xsys='1'; ysys='1'; x=100; y=100; output;

/* Annotate tick marks on line at top of graph */
function='move'; xsys='1'; ysys='1'; x=100*(0*(150/6))/150; y=100;
output;
function='draw'; ysys='7'; y=2; output;
function='cntl2txt'; output; function='label'; position='b';
text=trim(left(0*(150/6))); output;

function='move'; xsys='1'; ysys='1'; x=100*(1*(150/6))/150; y=100;
output;
function='draw'; ysys='7'; y=2; output;
function='cntl2txt'; output; function='label'; position='b';
text=trim(left(1*(150/6))); output;

{and so on, for the other tick marks}

/* Labels at top of the columns */
fuction='move'; xsys='3'; x=0; ysys='1'; y=100; text=''; output;
function='move'; ysys='9'; y=8; output;
function='label'; style="&font"; color="&graytext"; size=5.5;
ysys='3'; y=80;
xsys='3'; x=15; position='2'; text='Product'; output;
xsys='1'; x=45; position='2'; text='(Revenue, in $1000s)'; output;
xsys='3'; x=94; position='1'; text='Revenue'; output;

run;

data anno5; set anno5 anno5b;
run;
```

Create Plot 5:

Plotting the graph is similar to the others—use TITLE2 for a legend, run PROC GCHART, specify the ANNO= data set, and save the output in plot 5 to use later in PROC GREPLAY.

```
title1 height=2 " ";
title2 height=7 justify=left font=&fontb color=&browntext " Product
Sales YTD"
 justify=right height=5
   font=&font      color=&graytext "( "
   font="webdings" color="&revcolor" '67'x
```

```
    font=&font       color=&graytext " Revenue;    "
    font="webdings"  color="&unitcolor" '67'x
    font=&font       color=&graytext " Units;    "
    font=&font       color="black" '|'
    font=&font       color=&graytext "Target )    "
  ;
title3 height=9pct " ";
title4 angle=90 height=8pct " ";   /* blank space to left of bar chart
*/
title5 angle=-90 height=30pct " ";   /* blank space to right of bar
chart */

axis1 label=none order=('Cabernet' 'Zinfandel' 'Chardonnay' 'Sauvignan
Blanc' 'Merlot')
 value=(color=&backcolor justify=right font=&font  ) style=0
offset=(3,3);

axis2 label=('(Units sold, in 100s)') minor=none major=(number=6)
 value=(color=&graytext t=1 "0" t=2 "50" t=3 "100" t=4 "150" t=5 "200"
t=6 "250")
 color=&graytext offset=(0,0);

axis3 label=none value=none;

pattern1 v=solid color=&backcolor;

goptions xpixels=495 ypixels=240;
proc gchart data=data5 anno=anno5;
 hbar barvar / group=groupvar type=sum sumvar=normval
 gaxis=axis1 raxis=axis2 maxis=axis3
 width=1.2 space=0 gspace=1.7
 coutline=same nostats noframe
 name="plot5" ;
run;
```

Plot 5 Results:

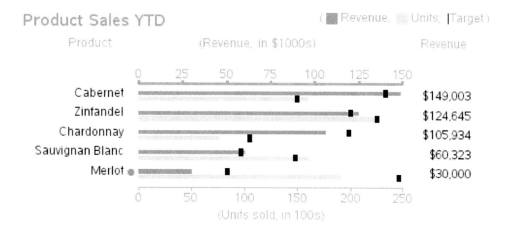

Plot 6 (Top 10 List)

Data for Plot 6:

Plot 6 is not really a plot at all. In the general sense it is a table. But the tool I am using to place all the graphs on the same page (PROC GREPLAY) works only with graphical output. I have to create this table using a "graph." Therefore, my tool of choice is a blank PROC GSLIDE with annotated text. The data is fairly simple—just the top 10 stores and their values.

```
data data6;
input customer $ 1-25 pipeline qtd ytd;
datalines;
The Big Wine Store        25585 33585 134777
Wines 'R Us               15865 12185  42984
Fruit of the Vine Inc.    15766 10766  39364
Spirits of the Age        11394 28865 120758
The Beverage Company      10624 16648  81834
Sips and Bites            10005  9764  24581
American Vintner's Best    8384 11865  38000
Barrel and Keg             6397 10766  39364
Cheers                     5394 22939  91056
Happy Hour                 4380  9551  18547
;
run;


proc sort data=data6 out=data6;
 by descending pipeline;
run;
```

Annotation for Plot 6:

Creating a graphical version of a table is a very useful thing. You might want to take special notice of this code. For each line of data, you output all the table values, each one at the desired column (X value). And the Y value is calculated based on the observation number of the data.

Note that, if desired, you can use an HTML= variable with HTML tags for data tips and drill-down links on each of the separate pieces of text that you annotate. You can also control the color of the annotated text and the background color behind the text.

```
/* Annotate the table values */
data anno6; set data6;
length text $100;
xsys='3'; ysys='3'; hsys='3';
rank=_n_;
function='label';
y=80-(rank*7);
x=5; position='4'; text=trim(left(rank)); output;
x=8; position='6'; text=trim(left(customer)); output;
x=60; position='4'; text=trim(left(put(pipeline,comma7.0))); output;
x=78; position='4'; text=trim(left(put(qtd,comma7.0))); output;
x=96; position='4'; text=trim(left(put(ytd,comma7.0))); output;
run;
```

```
/* Annotate the table column headers */
data anno6b;
length text $100;
xsys='3'; ysys='3'; hsys='3';
function='label';
color="&graytext";
y=82;
x=10; position='6'; text='Customer'; output;
x=60; position='4'; text='Pipeline'; output;
x=76; position='4'; text='QTD'; output;
x=94; position='4'; text='YTD'; output;
run;

data anno6; set anno6 anno6b;
run;
```

Create Plot 6:

Run PROC GSLIDE, specifying the annotate data set, and save the results in a GRSEG named PLOT6.

```
title1 height=8 justify=left font=&fontb color=&browntext
  "Top 10 Customers in the Pipeline (Revenue)";
footnote;

goptions xpixels=360 ypixels=240;
proc gslide name="plot6" anno=anno6;
run;
```

Plot 6 Results:

Top 10 Customers in the Pipeline (Revenue)

	Customer	Pipeline	QTD	YTD
1	The Big Wine Store	25,585	33,585	134,777
2	Wines 'R Us	15,865	12,185	42,984
3	Fruit of the Vine Inc.	15,766	10,766	39,364
4	Spirits of the Age	11,394	28,865	120,758
5	The Beverage Company	10,624	16,648	81,834
6	Sips and Bites	10,005	9,764	24,581
7	American Vintner's Best	8,384	11,865	38,000
8	Barrel and Keg	6,397	10,766	39,364
9	Cheers	5,394	22,939	91,056
10	Happy Hour	4,380	9,551	18,547

GREPLAY Template

This is where we pull it all together.

Using the size of the whole page, in combination with the sizes of the individual graphs, design a custom GREPLAY template to display the six graphs (PLOT1–PLOT6) and the title slide (TITLES) on the same page. When you create the graphs, you can name the GRSEGs anything you want, but I find it very useful to use a number in the name (such as 1–6) that corresponds to the numbered areas of the GREPLAY template.

See the other GREPLAY examples for more details about designing GREPLAY templates, and calculating the X and Y percentage values for each of the areas of the template.

The following code first defines the six areas of the custom template (areas 0–6), and then "replays" the seven GRSEGs (titles, and plot1-plot6) into the custom template.

```
proc greplay tc=tempcat nofs igout=work.gseg;
  tdef dashbrd des='Dashboard'

    0/llx = 0    lly =   0
      ulx = 0    uly = 100
      urx =100   ury = 100
      lrx =100   lry =   0

    1/llx = 0    lly = 60
      ulx = 0    uly = 90
      urx =55    ury = 90
      lrx =55    lry = 60

    2/llx =60    lly = 60
      ulx =60    uly = 90
      urx =100   ury = 90
      lrx =100   lry = 60

    3/llx = 0    lly = 30
      ulx = 0    uly = 60
      urx =55    ury = 60
      lrx =55    lry = 30

    4/llx =60    lly = 30
      ulx =60    uly = 60
      urx =100   ury = 60
      lrx =100   lry = 30

    5/llx = 0    lly =   0
      ulx = 0    uly = 30
      urx =55    ury = 30
      lrx =55    lry =   0

    6/llx =60    lly =   0
      ulx =60    uly = 30
      urx =100   ury = 30
      lrx =100   lry =   0
  ;
  template = dashbrd;
  treplay
      0:titles
    1:plot1 2:plot2
    3:plot3 4:plot4
```